Buying A Used Shortwave Receiver

A Market Guide to Modern Shortwave Radios

Fourth Edition   **By Fred Osterman**

Universal Radio Research
6830 Americana Parkway
Reynoldsburg, Ohio 43068
United States of America

©1998 by Fred Osterman
Universal Radio Research
Reynoldsburg, Ohio

FOURTH EDITION

FIRST PRINTING - November 1998

All rights reserved. Reproduction, publication, distribution, transmission or photocopying of all or any parts of this book, in any manner, without the prior written permission of the publisher is prohibited. While every precaution has been taken in the preparation of this book, the publisher assumes no responsibility for omissions or errors. No liability is assumed with respect to the use of the information herein. No patent liability is assumed with respect to the use of the information herein. All trademarks illustrated are the property of their respective companies.

International Standard Book Number: 1-882123-14-X
Library of Congress Catalog Card Number: 98-075112

Printed in the United States of America.

Contents

Preface 1

1 Introduction 3

2 Equipment Sources
   Private Sales 5
   Hamfests 6
   Radio Stores 10
   The Internet 11

3 More Information 12

4 Model Listings 17

5 Allied
   AX-190 24
   SX-190 24

6 AOR
   AR3030 25
   AR7030, AR7030+ 25

7 Drake
   R7 26
   R7A, R4245 26
   R8 27
   R8A 27
   R8B 28
   SPR-4 28
   SSR-1 29
   SW1, PRN1000 29
   SW2 30
   SW-4A, SW-4 30
   SW8 31

8 Grundig
- Satellit 400 32
- Satellit 500 32
- Satellit 650 33
- Satellit 700 33
- **YB-400,** YB-400PE 34
- **YB-500** 34

9 Icom
- **PCR-1000,** PCR1000-02, PCR-100 35
- R-70 35
- R-71A 36
- R-72 36

10 Japan Radio Company
- NRD-345 37
- NRD-505 37
- NRD-515 38
- NRD-525 38
- **NRD-535,** NRD-535D, NRD-535V 39
- NRD-545 39

11 Kenwood
- R-300 40
- R-600 40
- R-1000 41
- R-2000 41
- R-5000 42

12 Lowe
- HF-125 43
- **HF-150,** HF-150E, HF-150M 43
- **HF-225,** HF-225E 44
- **HF-250,** HF-250E 44

13 Magnavox
- D2935 45
- D2999 45

14 McKay Dymek
DR22C-6, DR22 46
DR33C-6 46
DR44, DR44-6 47
DR101-6 47

15 Panasonic
RF-799, RF-799LBS, RF-799LBE 48
RF-2600, DR26 48
RF-2800, DR28, RF-2900, DR29 49
RF-3100, DR31, RF-B30 49
RF-4800, DR48 50
RF-4900, DR49 50
RF-6300, RF-6300L, DR Q 63 51
RF-9000 51
RF-B40 52
RF-B45 52
RF-B60 53
RF-B65 53
RF-B300, DR-B300 54
RF-B600, DR-B600 54

16 Realistic (Radio Shack)
DX-100 55
DX-120 55
DX-150 56
DX-150A 56
DX-150B 57
DX-160 57
DX-200 58
DX-300 58
DX-302 59
DX-394 59

17 Sangean
ATS-803A, ATS-803, Realistic DX-440 60
ATS-808, ATS-808A, Realistic DX-380 60
ATS-818CS, ATS-818, Real. DX-390, DX-392 .. 61
ATS-909, Realistic DX-398 61

18 Sony
CRF-1 62
CRF-320A 62
CRF-330K 63
CRF-V21 63
ICF-2001 64
ICF-2002, ICF-SW7600D 64
ICF-2003, ICF-SW7600DS 65
ICF-2010, ICF-2001D 65
ICF-6500W, ICF-6500L 66
ICF-6700W, ICF-6700L 66
ICF-6800W, ICF-6800WA "Orange" ... 67
ICF-SW1S, ICF-SW1E 67
ICF-SW55 68
ICF-SW77 68
ICF-SW100S, ICF-SW100E 69
ICF-SW1000T, ICF-SW1000TS 69
ICF-SW7600 70
ICF-SW7600G, ICF-SW7600GS 70

19 Yaesu
FRG-7 71
FRG-100 71
FRG-7000 72
FRG-7700 72
FRG-8800 73

20 Other Manufacturers
Bearcat DX-1000 74
Galaxy R-530, R-1530 74
Racal RA6790/GM, RA6790 75
Ten-Tec RX-325 75
Uniden CR-2021, Realistic DX-400 ... 76
Watkins-Johnson HF1000 76

21 Model Index 77

Preface

Buying a used shortwave radio can be rewarding, but at the same time frustrating. Shortly after I joined Universal Radio, I became aware of the "frustrating" aspect. Universal, along with other radio dealers, would publish lists of used equipment giving manufacturer, model and price. It quickly became apparent that customers needed more information than this to make an informed purchase. Obtaining specifications on models *currently* in production has never been difficult. Current radio magazines and catalogs provide this. However, there were few sources for obtaining vital details on *discontinued* models. With this in mind, I published a book called *Shortwave Receivers Past & Present - First Edition*. This modest book covered portable and tabletop models produced in the past twenty years and was well received.

Many readers of this book felt that the next edition should include older radios. The Second Edition grew to include the numerous offerings from Hallicrafters, National, Hammarlund and many other famous manufacturers from the golden age of tube receivers.

Readers of the Second Edition were still not satisfied and urged that even more radios be added including additional European manufacturers, commercial grade sets and military models. As a result, the Third Edition expanded to nearly 500 pages covering 770 models, made from 1942 to 1997 from nearly one hundred different manufacturers.

Shortwave Receivers Past & Present - Third Edition has been accepted as the standard reference for collectors and admirers of shortwave communications receivers worldwide.

A new book titled *Portable Shortwave Receivers Past & Present* is in preparation to cover portable shortwave receivers in a similar fashion.

Both these books have evolved in size, scope and price beyond the needs of someone simply trying to purchase a modern used shortwave radio.

This book attempts to fill the *original* mission of the *First Edition* of *Shortwave Receivers Past and Present*. It is an affordable guide to modern (solid-state) shortwave receivers produced in the last twenty years. It covers the 100 most popular receivers used or sought by shortwave DXer's.

Chapter 3 - More Information suggests books that cover earlier models and less common radios. If your interest exceeds the scope of this book, please refer to these publications.

The author would like to thank David Crystal, Matt Stutterheim, John T. Wagner and James D. Piroli II for their help with this publication.

1 Introduction

This book is designed to provide the radio hobbyist with concise information on the value, features, specifications and performance of popular modern shortwave radios.

A key objective of this book is to help the reader make an informed choice when purchasing a modern used shortwave receiver. The following questions are addressed:

- When was the radio sold?
- What did it sell for when new?
- What is the current market value for the radio in used condition?
- Where was the radio made?
- What frequencies does it cover?
- What modes does it receive?
- What is the frequency display ... analog or digital?
- If a digital display, what is its resolution?
- What voltages (or batteries) are required?
- What accessories were made for it?
- How well did the receiver work when new?
- What is the used value rating of the receiver?

Some truly excellent values can be realized by purchasing a used shortwave radio. Receivers built in the last twenty years are solid state. Solid state means the circuit is composed of transistors and integrated circuits rather than vacuum tubes. The problems of tube aging, such as heat and wear, are not a factor with solid state sets. In fact, most solid state receivers work as well five or ten years after they were made as the day they came out of the factory. Therefore, purchasing a quality used solid state shortwave radio can afford substantial savings with no loss in performance.

Unlike computers, stereos and other consumer electronic items, shortwave receivers have a long product life. Shortwave receivers do not become obsolete after five or even ten years. In fact, a twenty year old receiver is probably more useful and usable than a three year old personal computer.

Why purchase a used receiver? The answer is simple: savings! You can typically expect to pay 30 to 40 percent off the price of a current model radio. On a discontinued model, the savings can be 50% or more.

In the era of vacuum tube receivers, purchasing a used receiver could be quite risky. As the tubes would age and become weaker, the performance of the receiver would gradually decline. The older tube-type sets would only work as well as the tubes in them. Tube receivers were also more mechanically intensive and would use either gears or strings for tuning the VFO. Band switching was accomplished by complex rotary switches. Older receivers required far more maintenance than today's solid-state digital radios. The performance of most solid-state (i.e. no vacuum tubes) radios does not diminish with time. A Yaesu FRG-100 that was built in 1992 will work as well as one purchased last week. This fact makes purchasing a used FRG-100 at a 30 to 40% savings quite tempting. With a little knowledge and preparation you can confidently purchase a used shortwave receiver.

This book does not include tube-era receivers. In most (but not all!) cases, they do not perform on par with today's solid-state rigs, and may pose potential parts and service difficulties. Most of the major manufacturers of tube radios (Hammarlund, Hallicrafters, etc.) are no longer in business. Today these radios are more sought by collectors than listeners. While a few models such as the Collins R-390/A, 51S-1 and Hammarlund HQ-180/A and SP-600 still perform admirably, most tube receivers may be best left to the collectors.

Portable *analog* shortwave models have not been included. There are only a handful of models that could even hope to compete in overall performance with today's digital shortwave portables. A few noteworthy examples would be the Barlow Wadley XCR-30 (left), Panasonic RF-2200 (right) or Sony ICF-7601. Again, unless there is collector or nostalgia interest, a quality *digital* portable would be the smarter choice for most.

2 Equipment Sources

■ Private Sales

You have done your technical research. You have limited your target radio down to two or three different models, and have also developed a good feel for what you should spend for each. Now it is time to find the radio.

One equipment source is other hobbyists. They have bought a new receiver and wish to sell their old one. Or perhaps they have decided to explore amateur radio, and are looking to raise capital for a transceiver. Still others may be leaving the hobby. Used receivers also occasionally show up at auctions and yard sales. Purchasing a radio (or most anything else) from a individual has advantages and disadvantages:

Advantages:

- ✦ When purchasing a radio from an individual you can normally expect to pay less than if you bought it from a radio store.

- ✦ An individual will not charge you sales tax.

Disadvantages:

- ✦ You must take the seller's word as to the history, condition, and performance level of the radio. If you are making the purchase in person, you can check the condition and performance level.

- ✦ If purchasing by mail, you cannot determine the integrity of the seller.

- ✦ You cannot determine if the radio is stolen property.

- ✦ An individual will not permit payment by credit card.

- ✦ An individual will not generally be in a position to offer a warranty.

■ Hamfests

Attending a hamfest can be a fun and educational experience. The flea-market at a major hamfest can be like a trip through a radio museum. It is an excellent opportunity to see many interesting and sometimes rare radios. Also be prepared to see an immense amount of unadulterated junk. But one man's garbage is another man's treasure!

A salesman at a leading East coast dealer suggests that you can never go wrong at a radio flea market ... as long as you take only $5.00 with you ... and that includes parking fees! That cautionary note may be an exaggeration, but one should be prepared.

When you enter the hamfest flea market, ask if a *test bench* has been set up. Most of the larger events provide a *test bench* as a service for those wishing to checkout a used equipment purchase. AC voltage (and 12 VDC) plus an antenna is provided. You can *fire-up* the radio and confirm that it is operational. Come prepared to test the radio, even if a test bench is not available.

■ A Very Long Extension Cord

You may not be able to find an outside AC outlet. With an extension cord, you can connect to an outlet near the door and bring the extension out.

■ A *Cheater* Antenna

Obtain a 15 to 20 foot length of plastic coated wire (18 or 20 gauge, preferably stranded). Connect a banana plug at one end and an alligator clip at the other.

With this configuration you will be able to add an antenna to virtually any radio. If the radio has a standard SO-239 jack (for a PL259-type plug) you can fit the banana plug end perfectly into the center of this jack. Use the alligator end to conveniently attach the antenna to any nonconductive item. If the receiver only has screw terminals, use the alligator end to attach to the radio. Such an antenna is not designed for DXing, but is adequate for receiver testing.

Testing a digital communications receiver will not take long, but should include the following considerations:

1. Look at the frequency display. Rotate the knob to ensure all the segments of all the digits work. The numeral **"8"** will test all segments.

2. Check to be sure all indicators are working. For example, if an LED is used to indicate mode, check all modes. Also be sure the lamp in the S-meter is working. The failure of an LED indicator or S-Meter lamp may not discourage you from buying the radio, but this information may help when you negotiate the best price.

3. Rotate the gain knob(s). If there is a *scratchy* sound, the potentiometer is dirty. If it is not serious, a "shot" of TV-tuner spray (without carbon tetrachloride) may clear it up. If it is very scratchy, the potentiometer may have to be replaced.

4. If there are rotary-switch knobs, try turning them. Do they snap in place and make a positive, clean contact or do they have a *mushy* feel and make a scratchy sound? Do you have to fiddle with the knob to help it make solid contact? Again, if the problem is minor, tuner spray can help. If the problem is advanced, the rotary switch would have to be replaced. This is a costly repair. If the *band switching* rotary knob has to be replaced, this is a very costly repair ... assuming the rotary switch can even be found.

5. Listen to the audio for signs of *hum.* A noticeable hum will indicate that the electrolytic capacitors in the receiver's power supply are going bad. This condition is especially prevalent in tube sets which are now reaching the age of twenty-plus years. Equipment that has been stored and not used for two or more years, will often suffer from capacitor failure. This condition can occur in radios bought from an estate sale.

Please note that some older tube radios with 600 ohm headphone output will exhibit some *hum* if used with low impedance headphones.

6. Now it is time to actually listen to the receiver for signals. You are probably listening to the radio during the day and probably with a modest antenna. As a shortwave listener, you already know that you will not be able to hear a lot of stations. However, you should hear some. Start with the 19 meter band 15000 to 15450 kHz. Look for WWV at 15000 kHz and broadcast stations above. Monitoring just one frequency is not an adequate test. It is important to remember that even general coverage digital receivers are *electronically* divided into bands. This is not normally apparent, but perhaps you have noticed the sensitivity of the receiver changes when going from 1600 to 1601 kHz; or perhaps on your receiver it might also change when going from 4000 to 4001 kHz. Within the range of shortwave (1600 to 30000 kHz), there may be five or six bands. Therefore, try to sample the radio over every 5 MHz. You should be able to at least hear background static at roughly the same level throughout the range of the receiver. If the last 5 or 6 MHz of the receiver are totally dead (not even hiss), the receiver may have been used by a CB'er to monitor his own signal. If the shortwave and CB antenna were too close, this may have damaged the highest *band* of the receiver.

If you are interested in purchasing the radio do not be shy about making an offer at substantially lower than the marked price (maybe even one half). If the seller is tired of carrying the radio around, he or she may let it go cheaply.

Advantages:

- ✦ When purchasing a radio at a hamfest or flea market, you normally will pay less than if you bought it from a radio store.

- ✦ At <u>most</u> hamfest flea markets you will not be asked to pay sales tax.

- ✦ At a flea market, you may be able to negotiate the price ... especially at the end of the day!

Disadvantages:

- You must often take the seller's word as to the history and performance level of the radio.

- You cannot determine if the radio is stolen property.

- An individual will not permit payment by credit card.

- An individual will not generally be in a position to offer a warranty.

Special Notes ...

● For safety reasons, very old, low-end, tube receivers without power transformers should not be plugged in until it can be established that they do not present a shock hazard. Such models should be examined by a qualified technician before power is applied to them.

If you are seeking an older tube-radio, bring mono headphones (with a ¼ inch plug), as some older radios did not have internal speakers.

■ Radio Stores

Most stores that sell new amateur and shortwave equipment also sell used equipment. Most have come in to the store when customers trade-up. Unfortunately, not every city has an amateur radio store. Buying used gear from an established and reputable radio store should not be a problem. The hobbyists radio publications *QST, CQ, Monitoring Times, Electric Radio, Radio Bygones* and *Popular Communications*, feature ads for radio stores. Many have a toll-free 800 number that you can call to check for a particular model, or request a used list. Some radio stores also post available used equipment on their Internet web sites.

Advantages:

✦ Most radio stores will thoroughly test a used receiver to be sure it meets its specifications before putting it out for sale. (Or they will mark it "as-is".)

✦ Most radio stores will offer a warranty including parts and labor. This is usually a 30 or 60 day limited warranty.

✦ Some radio stores will permit a return without penalty within seven to ten days.

✦ Most radio stores are available to answer any operational questions you may have.

✦ You may use a credit card.

Disadvantages:

✦ When purchasing a used radio at a radio store you will often pay more than at a flea market or private sale.

✦ If you actually purchase the radio in the store you will have to pay sales tax. If you are purchasing mail-order out of state, you may not have to pay sales tax in most cases.

A Reminder ...

When trading **in** a radio, expect to receive 20 to 25% less than the selling amount. Also understand that you will seldom fully recover the cost of special modifications and options when selling or trading a radio.

■ The Internet

The Internet is an increasingly popular place to find used and collector radios. All the above advantages and disadvantages of "Private Sales" listed above are applicable to the Internet ... plus the normal caveats of any business transaction over the Internet. Some related usenet news groups include:

```
rec.radio.amateur.Boatanchors
rec.radio.amateur.shortwave
rec.radio.amateur.swap
```

Some interesting mailing lists include:
```
WUN World Utility Network
VSS Vintage Solid State
Boatanchors
```

A few Web sites to try for information and/or actual equipment include:
```
http://www.antiqueradio.com
http://www.radiofinder.com
http://www.users.fast.net/~wa3key/collins.html
http://www.universal-radio.com
http://chide.bournemouth.ac.uk
http://www.rnw.nl/realradio/antique_index.html
http://www.torontosurplus.com/
http://www.mindspring.com/~johnmb/bawebpg.htm
http://cayman.ebay.com/aw/index.html
http://www.snafu.de/~wumpus/index.html
http://www.cyberventure.com/heathkit/ham
http://ouvaxa.cats.ohiou.edu/~post/Pix/BA.html
http://alpha.wcoil.com/~fairadio/
```

You can also use search engines like AltaVista to search for receiver manufacturers and sometimes even models.

3 More Information

It pays to be knowledgeable when selecting and purchasing a new or used radio. You must decide what performance level you need, what features you expect, and then try to match the budget. Obtaining information on new receivers is easy. Many radio stores offer informative catalogs and literature. Radio magazines show ads and even feature product reviews from time to time. Obtaining information on older receivers is a bigger challenge. The references that follow provide further technical information on current or older receivers and include models not covered in this book:

Shortwave Receivers Past & Present
Communications Receivers 1942-1997 - Third Edition
By Fred Osterman. Universal Radio Research ©1998 476 p
This huge guide provides vital information on over 770 communications receivers manufactured in the last 55 years. 108 Chapters with 840 photos.

Radio Receiver - Chance or Choice
by Rainer Lichte. Gilfer Associates, Inc., ©1985
Reviews many popular shortwave portables and table radios from the period 1975 to 1985. **More Radio Receiver - Chance or Choice** reviews 14 more radios produced in the mid 1980's.

WRTH Equipment Buyers Guide, 1993 Edition.
Billboard Publications.
Reviews many popular receivers. Includes reviews from several editions of the **World Radio TV Handbook.**

Passport To Worldband Radio
By Larry Magne. International Broadcast Services.
This annual publication is primarily a shortwave broadcast frequency - transmission directory. However, it also contains authoritative reviews on current portable and tabletop shortwave receivers.

Radio Netherlands Receiver Shopping List
by Jonathan Marks Radio Netherlands
An informative pamphlet providing information and evaluation of current portable and tabletop shortwave receivers.

Radios By Hallicrafters
by C. Dachis. Schiffer. Atglen, Pennsylvania. ©1996
A marvellous book with over 1000 sharp photos of radio receivers, transmitters, speakers, early TV sets and accessories from the famous Hallicrafters (and Ecophone) label. Technical descriptions of every known model including dates with new prices and current values.

World Radio TV Handbook
Billboard Publications.
This venerable annual publication primarily covers shortwave, medium wave, FM and television broadcasting. Most years also include equipment reviews.

Heathkit - A Guide to the Amateur Radio Products
by C. Penson WA7ZZE. Electric Radio Press, Inc., Durango, CO. ©1995
The definitive examination of all amateur, shortwave and related accessories offered by Heathkit. An outstanding compilation with large sharp photographs.

Communications Receivers. The Vacuum Tube Era-Fourth Edition
by R.S. Moore. R.S.M., Key Largo, Florida. ©1997
The golden age of vacuum tube receivers is revisited in this comprehensive book covering the period 1932 to 1981. Key facts, features and photographs of Breting, Collins, Echophone, Gonset, Hallicrafters, Howard, National, Hammarlund, Drake, Sargent, RME and many more.

I.B.S. Whitepaper Reviews
by Larry Magne
International Broadcast Services, Penn's Park, PA 18943

These authoritative reports, 12 to 20 pages in length, offer complete and unbiased evaluations and ratings on a single radio model. These exhaustive reviews are available for most current communications receivers (and some worldband portables). I.B.S. also publishes a special White Paper titled *How To Interpret Receiver Specifications and Lab Tests*. The I.B.S. White Paper reports can be obtained directly from the publisher and select radio dealers.

The following publications are valuable sources for one or more of the following: current equipment reviews, advertising of new equipment, classified advertising of used equipment, historical or restoration articles for collectors.

Antique Radio Classified ◆ Collecting and Restoration
P.O. Box 2
Carlisle, MA 01741

Amateur Radio Trader ◆ Amateur Radio Sales
P.O. Box 3729
Crossville, TN 38557

Collins Collectors Assoc. ◆ Collecting and Restoration
P.O. Box 840924
Pembroke Pines, FL 33084

Collins Journal ◆ Collecting and Restoration
David A. Knepper *W3BJZ*
Box 34
Sidman, PA 15955

CQ ◆ Amateur Radio
76 North Broadway
Hicksville, NY 11801

DX Ontario ◆ Shortwave DXing.
Ontario DX Association
P.O. Box 161, Station A
Willowdale, ON M2N 5S8
Canada

Eddystone Newsletter ◆ Collecting and Restoration
Eddystone User Group
Graeme Wormald *G3GGL*
Sabrina Dr.
Bewdley, Worcs. DY12 2RJ,
England

Electric Radio ◆ Collecting and Restoration
14643 County Road G
Cortez, CO 81321-9575

Ham Trader Yellow Sheets ◆ Amateur Radio Sales
P.O. Box 2057
Glen Ellyn, IL 60138

Hollow State Newsletter ◆ Collecting and Restoration
Ralph Sanserino
P.O. Box 1831
Perris, CA 92572-1831

Monitoring Times ◆ Radio Listening.
P.O. Box 98
Brasstown, NC 28902

The Journal ◆ Shortwave DXing.
N. American Shortwave Assoc.
45 Wildflower Rd.
Levittown, PA 19057

Popular Communications ◆ Radio Listening.
76 North Broadway
Hicksville, NY 11801

Practical Wireless ◆ Amateur Radio
PW Publishing Ltd.
Arrowsmith Court
Station Approach
Broadstone, Dorset BH18 8PW
England

QST ◆ Amateur Radio
American Radio Relay League
225 Main St.
Newington, CT 06111-9965

Radio Bygones　　　　　　◆ Collecting and Restoration
9 Wetherby Close
Broadstone
Dorset BH18 8JB
England

Radio Communications　　◆ Amateur Radio
Radio Society of Great Britain
Lambda House
Cranborne Road
Potters Bar, Herts, EN6 3JE
England

Shortwave Magazine　　　◆ Radio Listening
PW Publishing Ltd.
Arrowsmith Court
Station Approach
Broadstone, Dorset BH18 8PW
England

73 Amateur Radio　　　　◆ Amateur Radio
70 Route 202 N.
Peterborough, NH 03458

4 Model Listings

The information in the model listings include some abbreviations the reader will want to be familiar with. Many of these terms are further described later in this Chapter.

AGC	Automatic Gain Control
AM	Amplitude Modulation
BFO	Beat Frequency Oscillator
BITE	Built-in Testing Equipment
CW	Continuous Wave (Morse code)
DSP	Digital Signal Processing
DX	Distant X(trans)mitter
Fluor	Fluorescent
FM	Frequency Modulation
HF	High Frequency
HFO	High Frequency Oscillator
Hz	Hertz
IF	Intermediate Frequency
ISB	Independent Sideband
kHz	KiloHertz
LCD	Liquid Crystal Display
LED	Light Emitting Diode
LSB	Lower Sideband
LW	Longwave
MHz	MegaHertz
MW	Medium Wave
NB	Noise Blanker
NL	Noise Limiter
PBT	Pass Band Tuning
Phs	Phosphorescent
RF	Radio Frequency
RIT	Receive Incremental Tuning
SW	Shortwave
SWBC	Shortwave Broadcast
TCXO	Temperature Compensated Crystal Oscillator
VAC	Voltage Alternating Current
VDC	Voltage Direct Current
VFO	Variable Frequency Oscillator
VRIT	Variable Rate Incremental Tuning
USB	Upper Sideband
VHF	Very High Frequency

① **KENWOOD**
R-2000

② **General Coverage Communications Receiver**
③ **Made In:** Japan 1983-1992   ④ **Voltages:** 100/120/220/240 VAC
⑤ **Coverage:** 200-30000 kHz   ⑥ **Readout:** Digital Phos. 0.1
⑦ **Modes:** AM/FM/USB/LSB/CW   ⑧ **Selectivity:** 6/2.7/_ kHz
⑨ **Circuit:** Triple Conversion ⑩ **Physical:** 14.8x4.5x8.3" 12 Lbs.
❶ **Features:** ¼" Head. Jack, S-Meter, Attenuator, Tone, NB, Mute Terminals, Record Jack, Dual Clock-Timer, Tilt Bar, Dimmer, 10 Memories, Squelch, Record Activation, Dial Lock, Scan, Sweep, Beep Level Adjust, Carry Handle, Three Tuning Rates.
❷ **Accs.:** SP-100 Speaker, VC-10 Internal VHF Converter (118-174 MHz), YG-455C 500 Hz Filter, DCK-1 12 VDC Kit.
❸ **New Price:** $499-699   ❹ **Used Price:** $370-420   ❺ **Rating:** ★★★★
❻ **Comments:** A good value. The keys marked Ø through 9 are strictly for memory recall. They are not for frequency input.

① The manufacturer name and model number.
② Receiver type.
③ Country of manufacture and years marketed.
④ Voltage Requirements (also see Comments).
⑤ Frequency coverage in kHz. FM in this location indicates coverage of the FM broadcast band. FM[S] indicates FM stereo reception.
⑥ Type of frequency readout or display. For digital displays, the lowest display resolution for shortwave will be indicated in kHz.
⑦ Modes of reception. Note: FM does not mean FM broadcast band!
⑧ Selectivity. This will be indicated in kHz (at -6 dB) if known. In this example a 6 and a 2.7 kHz filter are supplied with an empty slot provided for an optional filter.
⑨ Circuit type. Superheterodyne, solid state unless specified otherwise.
⑩ Physical size in inches and pounds (Lbs.).
❶ The radio's features are indicated here (see the following page).
❷ Optional accessories.
❸ Selling price range when new (typically discount price to list price).
❹ Current used value (for good condition with manual).
❺ A rating of 1 to 5 stars for the model as *a value* on the used market.
★★★★★Excellent ★★★★Very Good ★★★Good ★★Fair ★Poor.
❻ Comments.

The following provides a brief explanation for the "Features" listed in this book:

¼" Head. Jack — Most tabletop receivers include a ¼" (6.3 mm) headphone jack. Most hobbyist receivers will disengage the internal speaker when the headphones are plugged in.

AFC — Automatic Frequency Control is a feature which compensates for variations in received signal frequency. Usually applies to FM broadcast band.

AGC — Automatic gain control is a feature that compensates for variations in received signal strength.

Antenna Trimmer — Older general coverage receivers typically employed a variable capacitor to roughly match the antenna to the receiver.

Attenuator — Under special circumstances, it may be desirable to reduce the sensitivity of the receiver.

AVC — Automatic volume control. See AGC.

Bandspread — A true bandspread system involves two calibrated dials. The main or band set dial is used to set the major range of frequencies. The bandspread dial spreads a small portion of the main dial over a wide range. This system provides for increased frequency accuracy but is predicated on the main dial having been properly set in order for the bandspread band to align itself.

Bandspread 0-100 — Less elaborate receivers may have a bandspread dial arbitrarily marked from 0 to 100. This system does not provide improved frequency accuracy. This arrangement is nothing more than a repeatable, electrical fine tuning.

BFO — The Beat Frequency Oscillator is circuit designed to clarify CW and/or single sideband signals.

BITE — BITE stands for Built-In Testing Equipment. Receivers with this feature have the ability to run a self-diagnostic test and determine the source of the problem, usually to the board level. This information is then shown on the receiver's display.

Calibrator	Crystal calibrators were frequently featured in receivers before the advent of digital displays. When turned on, the calibrator would generate a precise carrier, usually every 25 and/or 100 kHz.
CAT Jack	A Yaesu proprietary interface port to assist in the control of the receiver from a computer.
Carry Handle	A handle for the convenient moving of the radio.
Clock	An analog or digital clock.
Dial Drag Adjust	This mechanical adjustment permits the user to adjust how easily the tuning knob turns. This may be referred to as Dial Brake Adj.
Dial Lamp	One or more lamps that illuminate the radio dial from the back or from the side.
Dial Lamp Switch	A few receivers feature a switch to turn off the dial lamp. This extends battery life when the radio is not operating from the mains.
Dial Lock	This feature disables the tuning knob by physical or electrical means. This prevents the receiver from accidently being tuned off frequency.
Dimmer	The dimmer feature allows the user to adjust the intensity of the frequency display.
DSP	Digital signal processing.
Ferrite MW Ant.	Some table top receivers and most portables feature a ferrite bar antenna to improve medium wave (AM 540-1700 kHz) performance.
Fine Tuning	A mechanical or electrical adjustment to make small tuning adjustments.
IF Gain	A few receivers permit the IF gain to be adjusted, as well as the RF and AF gain.
IF Out Jack	A jack to allow access to the receiver's I.F. This may facilitate the connection of a spectrum display or other ancillary device.
IF Shift	An IF Shift control allows the listener to shift the IF passband of the receiver without changing the actual center frequency of the receiver. This control is useful when there is interference on one side of the signal.

ISB	Independent sideband is the transmission of separate intelligence on each sideband. A military station may send voice on lower side band and radioteletype on the upper side band. A feeder may use each sideband to send different audio.
Line Out Jack	This output jack provides a low level audio signal that is not affected by the volume control of the receiver. This fixed-level jack is typically used to drive a tape recorder or radioteletype decoder.
Keypad	A numeric keypad is used to quickly tune the receiver to the desired frequency.
Memories	Early memories or presets stored frequency information only. Other operational parameters such as mode and bandwidth were added later. Recently some receivers feature Alpha Memories which store the station name or callsign along the other parameters. This labeling feature is particularly useful when memory capacity exceeds 100 channels.
Memory Pass	A feature that lets you lock out a memory location from scanning.
Memory Scan	The receiver has the ability to scan the frequencies stored in memories. Also see Sweep.
Modular Construction	The receiver has some circuits built on plug-in printed circuit cards for easy service and/or replacement.
Mute Line	A jack or terminal that connects the receiver to a transmitter or transceiver. This connection will quiet or mute the receiver during times that the transmitter is transmitting.
Notch Filter	A notch filter suppresses a very narrow band of frequencies within the passband. An adjustable notch can be particularly effective at the reduction or elimination of heterodynes.
NB	A noise blanker is a device to reduce noise. It is usually most effective on man-made pulse type noise.

PBT	Pass band tuning permits the tuning of the passband without changing the receiver's frequency. It is very similar to IF Shift and is useful in reducing adjacent channel interference.
Preamp	A preamplifier applies an extra measure of amplification to the RF stage of the receiver.
Preselector	A preselector peaks the receiver circuits for maximum sensitivity on the frequency being received.
Record Out Jack	(See Line Out Jack)
Recorder Activation	Relay contacts that are usually normally open, that close when the timer turns on the radio. These contacts, when connected to the "remote" jack of a tape recorder, will start the tape recorder when the radio comes on.
RF Gain	The RF gain control adjusts the gain of the radio frequency circuits, thus controlling the sensitivity of the set.
RIT	Receive incremental tuning allows tuning plus or minus from the main VFO frequency. This is often used in amateur radio transceive situations.
RS-232	A jack or port on a receiver that permits communications and control of the receiver at RS-232 voltage levels with a terminal or computer.
Scan	As used in this book, Scan indicates the receiver can scan the memory channels. Also see Sweep.
S Indicator	Typically an LED or LCD bargraph that indicates signal strength.
S Meter	A meter that measures the relative strength of the incoming signal.
S/AF Meter	A meter that can indicate signal strength or audio output. A switch on the receiver selects which function of the meter is active.
Sensitivity	Same as RF Gain.
Speaker	A built-in speaker. Many early communications receivers did not feature a speaker built into the radio. It was thought that the vibration associated with the speaker would contribute to receiver instability or misalignment.

Speaker Switch	A switch to turn the speaker on or off.
Spinner Knob	An indent or protrusion of the main tuning knob to facilitate the rapid rotation of the knob.
Spkr. Terminals	Terminals for connection of an external speaker
Squelch	This control is used to eliminate unwanted background noise when monitoring an inactive frequency.
Sweep	Sweep indicates that the receiver can automatically tune the spectrum between two user defined frequency limits.
Sync. Detection	An amplitude modulated signal (such as used by medium wave and shortwave broadcast stations) consists of a carrier plus a lower and an upper sideband. Propagational fading may destabilize the carrier. Frequently one sideband may be distorted by a nearby interfering signal. Synchronous detection replaces the fading carrier with a pure carrier frequency with no level variation. The circuit then reconstitutes the signal from the stronger of the two sidebands. A stable signal with less distortion is the result.
TCXO	Temperature compensated crystal oscillator for improved stability.
Tilt Bar	A tilt bar or tilt feet that allow the receiver to be angled up from the desk for more convenient viewing.
Tilt Handle	A tilt handle has the same function as a tilt bar and additionally can be used as a carry handle.
Tone	Changes the balance between highs and lows on the audio output. In some cases this was a simple switch that cut the high tones.
VFO	A VFO or variable frequency oscillator is simply a tuner. Some receivers feature two tuners or VFOs within the set. This has a variety of uses.
VRIT Tuning	Variable Rate Incremental Tuning. With this feature, the faster the tuning control is spun, the higher is the rate of frequency change per revolution. This is also called Automatic Progressive Tuning Rate.

5 Allied

Allied Electronics
7410 Pebble Dr.
Ft. Worth, TX 76118

ALLIED AX-190

Amateur Band Communications Receiver
Made In:	Japan 1971-1973	**Voltages:**	110-120 VAC 12 VDC
Coverage:	Ham (See Comments).	**Readout:**	Analog Linear
Modes:	AM/LSB/USB-CW	**Selectivity:**	4 kHz
Circuit:	Double Conversion	**Physical:**	15x7x10" 20 Lbs.

Features: ¼" Head. Jack, S-Meter, Preselector, AGC, NL, Calibrator, AGC, Q-Multiplier, Record Jack, Mute Line, RF Gain, HFO Output.
Accs.: SP-190 Speaker.
New Price: $250   **Used Price:** $90-110   **Rating:** ★★★
Comments: Coverage: 3.5-4, 7-7.5, 14-14.5, 15-15.5, 21-21.5, 28-28.5, 28.5-29, 29-29.5, 29.5-30 MHz plus one optional 500 kHz band. This receiver requires a speaker. May be labeled as Realistic or Radio Shack AX-190.

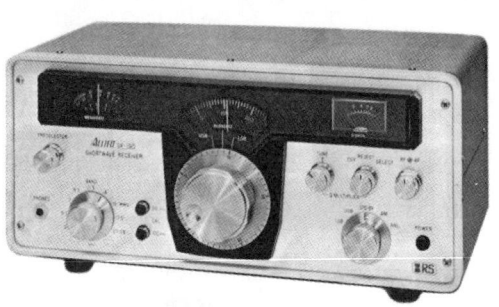

ALLIED SX-190

Shortwave Broadcast Communications Receiver
Made In:	Japan 1971-1973	**Voltages:**	110-120 VAC 12 VDC
Coverage:	SWBC (See Comments).	**Readout:**	Analog Linear
Modes:	AM/LSB/USB-CW	**Selectivity:**	4 kHz
Circuit:	Double Conversion	**Physical:**	15x7x10" 20 Lbs.

Features: ¼" Head. Jack, S-Meter, Preselector, AGC, NL, Calibrator, AGC, Q-Multiplier, Record Jack, Mute Line, RF Gain, HFO Output.
Accs.: SP-190 Speaker.
New Price: $250   **Used Price:** $100-140   **Rating:** ★★★
Comments: Coverage: 3.5-4, 5.7-6.2, 7-7.5, 9.5-10, 11.5-12, 14-14.5, 15-15.5, 17.5-18 and 27-27.5 MHz plus one optional 500 kHz band. This receiver requires a speaker. May be labeled as Realistic or Radio Shack SX-190. Please do not confuse the amateur band AX-190 with the SW band SX-190.

6 A.O.R.

A.O.R. Ltd.
2-6-4 Misuji, Taito-Ku
Tokyo 111, Japan

AR3030

General Coverage Communications Receiver
Made In:	Japan 1994-1997	**Voltages:**	120 VAC 11-16 VDC
Coverage:	30-30000 kHz	**Readout:**	Digital LCD 0.01
Modes:	AM/LSB/USB/FAX/FM	**Selectivity:**	6/2.4/_ kHz
Circuit:	Double Conversion	**Physical:**	10x3.55x9.5" 5 Lbs.

Features: ¼" Head. Jack, S-Meter, Sync. Detection, BFO, Keypad, RS232, Tone Switch, Dual VFOs, Dial Lamp, 100 Memories, AGC, ANL, Atten., Scan, Tilt Stand, IF Out., Memory Pass, Squelch, RF Gain.
Accs.: .5 kHz Filter, 2.5 kHz Filter, 4 kHz Filter, Internal VHF Converter.
New Price: $599-849   **Used Price:** $450-540   **Rating:** ★★★★
Comments: The AR3030 can also operate from eight AA cells stored internally. A nice, compact receiver.

AR7030

General Coverage Communications Receiver
Made In:	England 1996-1999	**Voltages:**	120 VAC 12-15 VDC
Coverage:	0-30000 kHz	**Readout:**	Digital LCD 0.01
Modes:	AM/LSB/USB/FAX/FM	**Selectivity:**	10/7/4.5/2.2/_/_ kHz
Circuit:	Double Conversion	**Physical:**	9.45x3.54x10" 5 Lbs.

Features: ¼" Head. Jack, S-Indicator, Synchronous Detection, BFO, Infrared Remote, PBT, Dial Lamp, 100 Memories, AGC, Attenuator, Tone, Squelch, Clock-Timer, Timer Contacts, Scan, LCD Contrast Adj., Record Jack, RS-232 Port, Tilt Stand, IF Gain.
Accs.: Various Filters, NB7030 CPU/Notch/NB, BP123 Battery, FL124 Daughter Board, TW7030 Telescopic Whip Antenna.
New Price: $999-1349   **Used Price:** $750-880   **Rating:** ★★★★★
Comments: This is a sophisticated, complex, highly configurable receiver. The model **AR7030+** features high tolerance components, CPU upgrade and 400 memories. The optional UPNB7030 upgrade adds NB/Notch to the AR7030+.

7 R.L. Drake Co.

R.L. Drake
230 Industrial Dr.
Franklin, OH 45005

DRAKE R7/DR7

General Coverage Communications Receiver

Made In:	United States 1978-1981	**Voltages:**	100/120/200/240 VAC
Coverage:	10-30000 kHz	**Readout:**	Digital LED 0.1
Modes:	AM/SSB/CW/RTTY	**Selectivity:**	2.3/_/_/_ kHz
Circuit:	Triple Conversion	**Physical:**	13.6x4.6x13" 19 Lbs.

Features: ¼" Head. Jack, S-Meter, PBT, Preamp, IF Notch, AGC, RIT, Ant. Switch, RF Gain, Calibrator, Counter Input Jack, Rec. Jack.
Accs.: MS-7 Speaker, NB-7A NB, AUX-7 Fixed Freq. Board, SL-300 300 Hz Filter, SL-500 500 Hz Filter, SL-1800 1.8 kHz Filter, SL-4000 4 kHz Filter, SL-6000 6 kHz Filter, RV-75 Ext. VFO.
New Price: $1295-1549   **Used Price:** $650-750   **Rating:** ★★★★
Comments: A favorite among tropical band DXers. A bit cumbersome to tune.

DRAKE R7A

General Coverage Communications Receiver

Made In:	United States 1981-1983	**Voltages:**	100/120/200/240 VAC
Coverage:	10-30000 kHz	**Readout:**	Digital LED 0.1
Modes:	AM/SSB/CW/RTTY	**Selectivity:**	9/2.3/.5/_/ kHz
Circuit:	Triple Conversion	**Physical:**	13.6x4.6x13" 19 Lbs.

Features: ¼" Head. Jack, S-Meter, PBT, Preamp, IF Notch, AGC, RIT, NB, Ant. Switch, RF Gain, Calibrator, Counter Input Jack, Rec. Jack.
Accs.: MS-7 Speaker, AUX-7 Fixed Freq. Board, SL-300 300 Hz Filter, SL-1800 1.8 kHz Filter, SL-4000 4 kHz Filter, SL-6000 6 kHz Filter, RV-75 External Synthesized VFO.
New Price: $1489-1649   **Used Price:** $690-840   **Rating:** ★★★★
Comments: Similar to the R7, but with the 500 Hz filter and NB-7A NB. The scarce and desirable optional RV-75 improved stability ($380 used). The very scarce **R4245** military-commercial R7A version has full frequency synthesis.

DRAKE
R8

General Coverage Communications Receiver
Made In: United States 1991-1995 **Voltages:** 100/120/200/240 VAC
Coverage: 10-30000 kHz **Readout:** Digital LCD 0.1
Modes: AM/SSB/CW/RTTY/FM **Selectivity:** 6/4/2.3/1.8/.5 kHz
Circuit: Double Conversion **Physical:** 13.2x5.25x13" 13 Lbs.
Features: ¼" Head. Jack, S-Meter, PBT, Preamp, IF Notch, AGC, BFO, 100 Memories, Sync. Detection, Tone, Dual Clock Timers, RF Gain, Recorder Activation, RS232 Port, Scan, Sweep, Keypad, Squelch, Dual VFOs, Speaker Switch, Line Output, Attenuator, Mute. The LCD display on the R8/A/B is nicely backlit.
Accs.: MS-8 Speaker, VHF Converter.
New Price: $980-999   **Used Price:** $650-720   **Rating:** ★★★★★
Comments: Drake's first shortwave receiver in ten years. A nice unit. The optional internal VHF converter covers 35-55 and 108-174 MHz.

DRAKE
R8A

General Coverage Communications Receiver
Made In: United States 1995-1998 **Voltages:** 100/120/200/240 VAC
Coverage: 10-30000 kHz **Readout:** Digital LCD 0.1
Modes: AM/SSB/CW/RTTY/FM **Selectivity:** 6/4/2.3/1.8/.5 kHz
Circuit: Double Conversion **Physical:** 13.2x5.25x13" 13 Lbs.
Features: ¼" Head. Jack, S-Meter, PBT, Preamp, IF Notch, AGC, BFO, 440 Alpha Mems., Sync. Detection, Tone, Dual Clock Timers, RF Gain, Recorder Activation, RS232 Port, Scan, Sweep, Keypad, Squelch, Dual VFOs, Speaker Switch, Line Output, Attenuator.
Accs.: MS-8 Speaker, VHF Converter.
New Price: $1050-1099 **Used Price:** $730-810   **Rating:** ★★★★★
Comments: Many ergonomic and performance improvements were added including: alphanumeric memories, faster scanning, improved AGC, improved notch, improved display, easier mode and bandwidth selection, tilt-bar, enhanced tone control, detachable line cord & expanded RS-232 command set.

DRAKE
R8B

General Coverage Communications Receiver
Made In: United States 1997-1999 **Voltages:** 100/120/200/240 VAC
Coverage: 100-30000 kHz[1] **Readout:** Digital LCD 0.1
Modes: AM/SSB/CW/RTTY/FM **Selectivity:** 6/4/2.3/1.8/.5 kHz
Circuit: Double Conversion **Physical:** 13.2x5.25x13" 13 Lbs.
Features: ¼" Head. Jack, S-Meter, PBT, Preamp, IF Notch, AGC, BFO, 1000 Alpha Mems., Sync. Detect., Tone, Dual Clock Timers, RF Gain, Recorder Activation, RS232 Port, Scan, Sweep, Keypad, Squelch, Dual VFOs, Speaker Switch, Line Output, Attenuator.
Accs.: MS-8 Speaker, VHF Converter.
New Price: $1159-1199 **Used Price:** $850-860 **Rating:** ★★★★★
Comments: Ergonomic and performance improvements over the R8A include: faster scanning, sideband selectable synchronous AM detection and 1000 alphanumeric memories. [1]Later production units tune from 10 to 30000 kHz.

DRAKE
SPR-4

Shortwave Broadcast Band Communications Receiver
Made In: United States 1969-1978 **Voltages:** 120/240 VAC 12 VDC
Coverage: See Comments. **Readout:** Analog Linear
Modes: AM/LSB/USB/CW **Selectivity:** 4.8/2.4/.4 kHz
Circuit: Double Conversion **Physical:** 10.75x5.5x12.25" 14 Lbs.
Features: ¼" Head. Jack, S-Meter, IF Notch, Preselector, RF Gain, Mute, Dial Lamp, Dial Lamp Switch, Spinner Knob, Audio In-Out Jack, Anti-Vox Jack.
Accs.: AL-4 LW/MW Loop, MS-4 Spkr, 5NB NB, SCC-4 Cal, FS-4 Synthz.
New Price: $379-699 **Used Price:** $250-360 **Rating:** ★★★
Comments: Ranges: .15-.5, .5-1, 1-1.5, 6-6.5, 7-7.5, 9.5-10, 11.5-12, 15-15.5, 17.5-18 and 21.5-22 MHz plus 14 additional 500 kHz crystal positions. Dial accuracy is ±1 kHz. Drake's first completely solid-state receiver.

DRAKE
SSR-1

General Coverage Communications Receiver
Made In: Japan 1975-1978 **Voltages:** 117/240 VAC 8 x D
Coverage: 500-30000 kHz **Readout:** Analog Linear
Modes: AM/LSB/USB/CW **Selectivity:** 5.5/3 kHz
Circuit: Double Conversion **Physical:** 13x5.5x11" 14 Lbs.
Features: ¼" Head. Jack, S-Meter, Preselector, Clarifier, Mute, Dial Lamp, Dial Lamp Switch, Spinner Knob.
Accs.: DC-PC 12 VDC Cord.
New Price: $299-350   **Used Price:** $130-200   **Rating:** ★★
Comments: The MHz knob (top left) tunes inner MHz dial on display. The main tuning knob (large) rotates the outer dial (10 kHz graduations). If the AC line voltage fails an automatic circuit switches to the internal D cells. This was the only Drake receiver made outside of Ohio.

DRAKE
SW1

General Coverage Broadcast Receiver
Made In: United States 1996-1999 **Voltages:** 12 VDC or 120 VAC
Coverage: 100-30000 kHz **Readout:** Digital LED 1.
Modes: AM **Selectivity:** 5.5 kHz
Circuit: Double Conversion **Physical:** 10.79x4.4x7.7" 5 Lbs.
Features: Mini Head. Jack, RF Gain, 32 Memories, Dimmer, Two Tuning Speeds, Keypad, Up-Down Tuning, Manual Tuning.
Accs.: Carry Handle, Mobile Mounting Bracket.
New Price: $200-300   **Used Price:** $140-145   **Rating:** ★★★
Comments: An affordable, basic, broadcast receiver without SSB capability. Supplied with an 120 VAC adapter. The earlier O.E.M. model **PRN1000** (produced for the "People's Radio Network") is similar, but with only one memory, a Local-DX switch and a Tone control in place of the RF Gain.

DRAKE

SW2

General Coverage Communications Receiver
Made In: United States 1996-1999 **Voltages:** 12 VDC or 120 VAC
Coverage: 100-30000 kHz **Readout:** Digital LED 1.
Modes: AM/LSB/USB **Selectivity:** 9/2.3 kHz
Circuit: Double Conversion **Physical:** 10.79x4.4x7.7" 6 Lbs.
Features: Mini Head. Jack, RF Gain, 100 Memories, Dimmer, Two Tuning Speeds, Keypad, Up-Down Tuning, Manual Tuning, External Speaker Jack, Synchronous Detection.
Accs.: Carry Handle, MMK-1 Mobile Mounting Bracket, Infrared Remote Control (shown).
New Price: $490-500   **Used Price:** $290-310   **Rating:** ★★★★★
Comments: Please note that the Remote Control (shown) is optional. The Drake SW-2 has nice audio and tunes in 50 Hz steps for easy SSB reception.

DRAKE

SW-4A

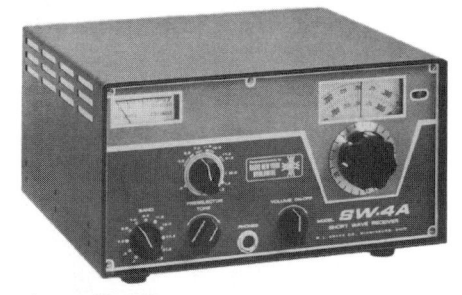

Shortwave Broadcast Band Broadcast Receiver
Made In: United States 1967-1974 **Voltages:** 120/240 VAC
Coverage: See Comments. **Readout:** Analog Linear
Modes: AM **Selectivity:** 5 kHz
Circuit: Double Conversion Hybrid **Physical:** 10.75x5.5x12.25"16Lbs.
Features: ¼" Head. Jack, S-Meter, Preselector, Dial Adjust, Tone, Dial Lamp.
New Price: $289-335   **Used Price:** $250-350   **Rating:** ★★★
Comments: Ranges: .15-.5, .485-1.05, .95-1.55, 5.95-6.55, 6.95-7.55, 9.45-10.05, 11.45-12.05, 14.95-15.45, 17.45-18.05, 21.45-22.05 and 25.45-26.05 MHz   Tuning this receiver is quick and easy. Dial accuracy is ±3 kHz. Note that this model does not receive CW/SSB. The earlier, very scarce **SW-4** (1966) is also a hybrid with eight tubes plus semiconductors and has a band switch *and* a range switch plus three 500 kHz auxiliary crystal sockets.

DRAKE
SW8

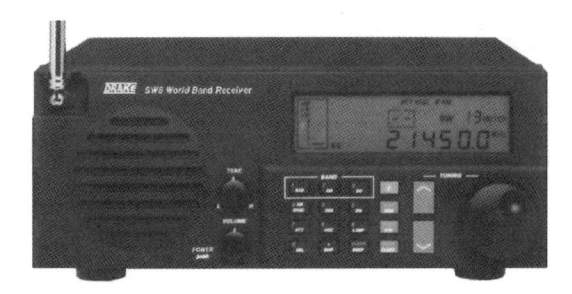

General Coverage Portable Communications Receiver
Made In: United States 1994-1999   **Voltages:**   12 VDC or 120 VAC
Coverage: 500-30000 kHz +FM +Air   **Readout:**   Digital LCD 1.
Modes: AM/SSB/CW   **Selectivity:** 6/4/2.3 kHz
Circuit: Double Conversion   **Physical:**   11.5x5.25x13" 10 Lbs.
Features: Mini Head. Jack, S-Indicator, RF Gain, 70 Memories, Dimmer, BFO, Memory Scan, Dual Clock Timer, Telescopic Whip, Whip Preamp, Two Tuning Speeds, Synchronous Detection, Keypad, Tone, Squelch (VHF AIR), Backlit Display, AGC.
Accs.:   MS-8 Speaker, Canvas Carry Case.
New Price: $589-699   **Used Price:** $350-435   **Rating:** ★★★★★
Comments: Note the coverage of the FM broadcast band 88-108 MHz and 118-137 MHz VHF aeronautical band. Units produced prior to March 1995 do not have the whip preamp. Units produced after January 1996 add longwave coverage tuning from 100-30000 kHz.

8 Grundig

Grundig
P.O. Box 2307
Menlo Park, CA 94026

GRUNDIG
Satellit 400

General Coverage Portable Communications Receiver
Made In: Portugal 1986-1989 **Voltages:** 110-127/220-240 VAC
Coverage: 148-30000 kHz + FM **Readout:** Digital LCD 1.
Modes: AM/SSB **Selectivity:** Two Positions
Circuit: Double Conversion **Physical:** 11.8x7x2" 5 Lbs.
Features: Mini Head. Jack, S/Battery-Meter, Treble, Bass, Keypad, Clock, 24 Memories, Sleep, Lock, Dial Lamp, Dial Lamp Switch. Memory Scan, Carry Handle, BFO, Antenna Jack.
New Price: $400-450 **Used Price:** $140-150 **Rating:** ★★★★
Comments: Also operates from six D cells and three AA cells or 12 VDC. Features a beefy six watt audio amplifier.

GRUNDIG
Satellit 500

General Coverage Portable Communications Receiver
Made In: Portugal 1990-1993 **Voltages:** 110-127/220-240 VAC
Coverage: 140-30000 kHz + FM[S] **Readout:** Digital LCD 1.
Modes: AM/SSB **Selectivity:** Two Positions.
Circuit: Double Conversion **Physical:** 12.25x7.25x3" 4 Lbs.
Features: Mini Head. Jack, S/Battery-Meter, Treble, Bass, Keypad, 42 Alpha Memories, Local-DX Switch, Sleep, Carry Handle, Dial Lamp, Scan, Record Jack, Record Activation, BFO, Lock, Clock-Timer, Synchronous Detection, 9/10 kHz MW, Ant. Jack.
New Price: $389-499 **Used Price:** $210-230 **Rating:** ★★★★★
Comments: One of the first portable receivers with inexpensive high quality IF filters. Selectivity is good. Also operates from four D cells or 12 VDC.

GRUNDIG
Satellit 650

General Coverage Portable Communications Receiver
Made In: Portugal 1987-1990   **Voltages:** 110-127/220-240 VAC
Coverage: 140-30000 kHz + FM^S **Readout:** Digital LCD 1.
Modes: AM/LSB/USB   **Selectivity:** Three Bandwidths.
Circuit: Double Conversion   **Physical:** 19.8x9.5x8" 19 Lbs.
Features: ¼" Head. Jack, S/Battery-Meter, Treble, Bass, Keypad, ANL, 60 Memories, Wide-Medium-Narrow, Sleep, FM Stereo, Lock, Carry Handle, Antenna Jack, Dial Lamp, Clock 24 Hr., Local-DX, Motorized Auto-Tune Preselector, 9/10 kHz MW Step.
Accs.: ACC476 NiCad Battery Pack
New Price: $900-1050   **Used Price:** $590-650   **Rating:** ★★★★
Comments: Also operates from six D cells (and two AA cells) or 12 VDC. 32 of the 60 memories are available for shortwave. Exceptionally full audio fidelity from the 15 watt amplifier. This model has a strong following.

GRUNDIG
Satellit 700

General Coverage Portable Communications Receiver
Made In: Portugal 1992-1996   **Voltages:** 110-127/220-240 VAC
Coverage: 140-30000 kHz + FM^S **Readout:** Digital LCD 1.
Modes: AM/LSB/USB   **Selectivity:** Two bandwidths.
Circuit: Double Conversion   **Physical:** 12.25x7.25x3" 4 Lbs.
Features: ¼" Head. Jack, S/Battery-Meter, Treble, Bass, Keypad, Clock, 512 Alpha Memories, Dial Lamp, Sleep, FM Stereo, Lock, AGC, RDS, 9-10 kHz MW Step, DX-Local Switch, Carry Handle, Synchronous Detect., Record Jack, Record Activation, Ant. Jack.
Accs.: Memory EPROM ICs
New Price: $479-599   **Used Price:** $230-250   **Rating:** ★★★★★
Comments: Also operates from four D cells or 12 VDC input. Three additional EPROM sockets are featured; each capable of holding a 512 preset integrated circuit (for a maximum of 2048 memories). An outstanding receiver.

GRUNDIG
YB-400

General Coverage Portable Communications Receiver
Made In: China 1993-1998 **Voltages:** 9 VDC 6xAA
Coverage: 140-30000 kHz + FM[S] **Readout:** Digital LCD 1.
Modes: AM/SSB **Selectivity:** Two Position
Circuit: Double Conversion **Physical:** 8x5x1.5" 1.4 Lbs.
Features: Mini Head. Jack, S-Indicator, Batt-Indicator, Tone Switch, 40 Memories, Dial Lamp, Sleep, FM Stereo, Lock, Local-DX, Keypad, Ant. Jack, 1/9/10 kHz MW Step, Clock-Timer 24 Hr.
Accs.: 400ACA AC Adapter.
New Price: $170-250   **Used Price:** $90-110   **Rating:** ★★★★★
Comments: Also called the Yacht Boy 400. The YB-400 is supplied with vinyl carrying case, wind-up antenna and stereo earplugs. Model **YB-400 PE** "Professional Edition" features a titanium colored case and is supplied with the Grundig AC adapter ($200-270 new). Both units have enjoyed wide acceptance.

GRUNDIG
YB-500

General Coverage Portable Communications Receiver
Made In: Portugal 1993-1997 **Voltages:** 110-127/220-240 VAC
Coverage: 150-30000 kHz + FM[S] **Readout:** Digital LCD
Modes: AM/LSB/USB **Selectivity:** 6 kHz
Circuit: Double Conversion **Physical:** 4.4x7.5x1.6"
Features: Mini Head. Jack, S Indicator, Dial Lamp, Dial Lamp Switch, 40 Memories, RDS FM, 90 ROM Frequencies, Tilt Stand, Keypad, Lock, Dual 24 Hour Clocks, Up-Down Tuning, Tone Switch, Line Output, Recorder Activation.
New Price: $299-349   **Used Price:** $119-139   **Rating:** ★★★
Comments: Also called the Yacht Boy 500. Supplied with an AC adapter or operates from four AA cells or 9 VDC. The YB-500 is an attractive unit, but actually falls short of the more affordable YB-400, in terms of performance.

9 ICOM

Icom America, Inc.
2380 116th Ave. NE
Bellevue, WA 98004

ICOM PCR1000

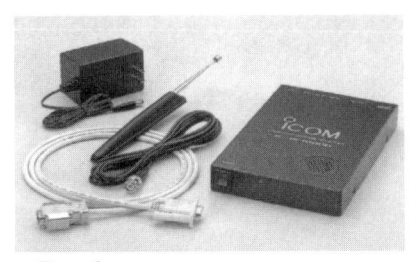

Wideband Computer Communications Receiver

Made In:	Japan 1997-1999	Voltages:	12.8 VDC 0.7A
Coverage:	100 kHz-1.3 GHz	Readout:	PC Computer 0.001
Modes:	AM/SSB/CW/WFM/NFM	Selectivity:	15/6/2.8 kHz
Circuit:	Double Conversion	Physical:	5x7.9x1.2" 2.2 Lbs.

Features: S-Meter Indicator, Unlimited Memories, Voice Scan, IF Shift, Clock, AGC, Packet Jack, Band Scope ±200 kHz, Attenuator, Scan, Sweep, NB, Speaker Jack, Squelch, Unlimited Memories.
Accs.: UT-106 DSP Option, OPC-131 DC Cord.
New Price: $400-600   Used Price: $300-340   Rating: ★★★★
Comments: Requires a PC with at least a 486x4 CPU and Windows 3.1. The American version PCR1000-02 has cellular frequencies blocked. Simpler model **PCR-100** (late 1998) does not include SSB or real-time band scope.

ICOM R-70

General Coverage Communications Receiver

Made In:	Japan 1982-1984	Voltages:	100/117/220-240 VAC
Coverage:	100-30000 kHz	Readout:	Digital Fluor. 0.1
Modes:	AM/SSB/CW/RTTY	Selectivity:	6/2.3/_ kHz
Circuit:	Quadruple Conversion	Physical:	11.3x4.4x10.9" 16 Lbs.

Features: ¼" Head. Jack, S-Meter, PBT, Preamp, IF Notch, AGC, BFO, Attenuator, Tone, Preamp, Dual VFOs, Squelch, Dimmer, NB, Scope Jack, Monitor, Line Output Jack, Dial Lock, RIT, Converter Input Jack.
Accs.: FL-63 CW Filter, FL-44 SSB Filter, EX-257 FM Mode Option, SP-3 Speaker.
New Price: $600-700   Used Price: $370-390   Rating: ★★★★
Comments: A very good receiver, especially for CW and SSB, but a bit quirky and complex to operate, especially at the edges of the 1 MHz bands.

ICOM
R-71A

General Coverage Communications Receiver

Made In:	Japan 1984-1996	**Voltages:**	100/117/220-240 VAC
Coverage:	100-30000 kHz	**Readout:**	Digital Fluor. 0.1
Modes:	AM/SSB/CW/RTTY	**Selectivity:**	6/2.3/_ /_ kHz
Circuit:	Quadruple Conversion	**Physical:**	11.8x4.4x10.9" 17 Lbs.

Features: ¼" Head. Jack, S-Meter, PBT[1], Preamp, IF Notch, AGC, BFO, Tone, Preamp, Dual VFOs, Squelch, Dimmer, 32 Memories, Attenuator, Squelch, Keypad, Dial Tension Adjust, Lock.

Accs.: FL-63 CW Filter, FL-44A SSB Filter, EX-257 FM Mode Option, CR-64 High Stability, EX-309 Intf. Option, CT-17 Level Converter, UX-14 CI-IV/CI-V Conv., MB-12 Mob. Bracket, EX-310 Voice Syn.

New Price: $689-1280   **Used Price:** $560-670   **Rating:** ★★★★★

Comments: An excellent receiver. [1]Some production did not include the PBT.

ICOM
R-72

General Coverage Communications Receiver

Made In:	Japan 1990-1998	**Voltages:**	100/117/220-240 VAC
Coverage:	100-30000 kHz	**Readout:**	Digital LCD 0.01
Modes:	AM/SSB/CW	**Selectivity:**	6/2.3/_ kHz
Circuit:	Double Conversion	**Physical:**	9.5x3.7x9" 10.6 Lbs.

Features: ¼" Head. Jack, S-Meter, Preamp, AGC, Dimmer, 99 Memories, Tone, Preamp, Squelch, Attenuator, Keypad, Lock, Tilt Bar, Dial Tension Adjust, Scan, Sweep, 24 Hour Clock-Timer, Three Tuning Rates.

Accs.: FL-100 500 CW Filter, FL-101 CW 250 Filter, UR-1 Protector, UT-36 Voice Synthz., UI-8 FM Mode, CT-17 Level Converter, CR-64 High Stability, OPC-131 DC Kit, MB-5 Mobile Bracket.

New Price: $696-1099   **Used Price:** $400-470   **Rating:** ★★★

Comments: Units made after June 1996 may not include the AC adapter.

10 Japan Radio Co.

Japan Radio Co. Ltd.
1011 SW Klickitat Way #B100
Seattle, WA 98134

[JRC] *Japan Radio Co., Ltd.*

NRD-345

General Coverage Communications Receiver
Made In: Japan 1997-1999 **Voltages:** 100/120/220/240 VAC
Coverage: 100-30000 kHz **Readout:** Digital LCD 0.01
Modes: AM/LSB/USB/CW/FAX **Selectivity:** 4/2/_ kHz
Circuit: Double Conversion **Physical:** 9.9x4x9.4" 7.8 Lbs.
Features: ¼" Head. Jack, S-Meter, Attenuator, NB, Mute Terminals, NB, Line Out Jack, BFO, AGC, Speaker Output, Keypad, Tone, 100 Memories, Scan, Sweep, Sync. Detection and RF Gain.
Accs.: CFL-231 300 Hz Filter, CFL-232 500 Hz Filter, CFQ-8673 Filter Board, CFL-233 1000 Hz Filter, 6ZCJD00350 RS-232 Cable.
New Price: $799-899 **Used Price:** $500-530 **Rating:** ★★★★★
Comments: Memories store: frequency, VFO, mode, bandwidth, NB, AGC and ATT settings.

[JRC] *Japan Radio Co., Ltd.*

NRD-505

General Coverage Communications Receiver
Made In: Japan 1977-1979 **Voltages:** 100/115/200/230 VAC
Coverage: 100-34000 kHz **Readout:** Digital LED 0.1
Modes: AM/USB/LSB/CW/RTTY **Selectivity:** 4.4/2/_ kHz
Circuit: Double Conversion **Physical:** 13.4x5.5x11.8" 22 Lbs.
Features: ¼" Head. Jack, S-Meter, Attenuator, NB, Mute Terminals, NB, RF Gain, Line Out Jack, RIT, BFO, AGC, Speaker Output, IF Out Jack, Line Output Jack, Modular Design.
Accs.: CW Filter 600 Hz, NVA-505 Speaker, CGA-26 Transceive VFO Converter, CD4-8 Four Channel Internal Memory Unit.
New Price: $2250-2275 **Used Price:** $790-850 **Rating:** ★★★★
Comments: Built to the highest physical standards. This model requires a speaker. Very scarce.

[JRC] *Japan Radio Co., Ltd.*
NRD-515

General Coverage Communications Receiver
Made In:	Japan 1979-1986	**Voltages:**	100/117/220/240 VAC
Coverage:	100-34000 kHz	**Readout:**	Digital LED 0.1
Modes:	AM/USB/LSB/CW/RTTY	**Selectivity:**	6/2.4/_/_ kHz
Circuit:	Double Conversion	**Physical:**	13.4x5.5x11.8" 17 Lbs.

Features: ¼" Head. Jack, S-Meter, Attenuator, NB, Mute Terminals, ANL, Line Out Jack, RIT, BFO, AGC, PBT, Speaker Output, MW Tune, Up-Down Tuning, Dial Lock, RF Gain, Int./Ext. VFO Switch.
Accs.: CFL-230 300 Hz Filter, CFL-260 600 Hz Filter, NVA-515 Spkr., CFL-218 1800 Hz Filter, NCM-515 Wired Keypad, NDH-515 24 Ch. Ext. Memory, NDH-518 96 Ch. Ext. Memory.
New Price: $900-1400   **Used Price:** $640-680   **Rating:** ★★★★★
Comments: A robustly built, top quality receiver. Optional memory units store frequency only. This model requires a speaker.

[JRC] *Japan Radio Co., Ltd.*
NRD-525

General Coverage Communications Receiver
Made In:	Japan 1986-1992	**Voltages:**	100/120/220/240 VAC
Coverage:	90-34000 kHz	**Readout:**	Digital Fluor. 0.01
Modes:	AM/SSB/CW/RTTY/FAX/FM	**Selectivity:**	4/2/_/_ kHz
Circuit:	Double Conversion	**Physical:**	13.5x5.2x11.3" 19 Lbs.

Features: ¼" Head. Jack, S-Meter, Attenuator, NB, Mute Terminals, ANL, Line Out Jack, RIT, BFO, AGC, PBT, Speaker Output, Keypad, 200 Memories, Up-Down Tuning, Tone, RF Gain.
Accs.: CFL-231 300 Hz Filter, CFL-232 500 Hz Filter, NVA-88 Speaker, CFL-233 1000 Hz Filter, CFL-218A 1800 Hz Filter, CMH-530 RTTY Demod., CMH-532 RS-232 Interface, CMK-165 Internal VHF-UHF Converter, 6ZCJD00139 Printer Cable for CMH-530.
New Price: $999-1249   **Used Price:** $520-550   **Rating:** ★★★★★
Comments: Memories store: frequency, mode, bandwidth, AGC and ATT.

JRC *Japan Radio Co., Ltd.*
NRD-535

General Coverage Communications Receiver

Made In:	Japan 1991-1998	**Voltages:**	100/120/220/240 VAC
Coverage:	100-3000 kHz	**Readout:**	Digital Fluor. 0.01
Modes:	AM/SSB/CW/RTTY/FAX/FM	**Selectivity:**	4/2/_/_ kHz [12 FM]
Circuit:	Triple Conversion	**Physical:**	13x5.2x11.3" 20 Lbs.

Features: ¼" Head. Jack, S-Indicator, Atten., NB, Mute, ANL, Notch, PBT, Line Out, Scan, Sweep, RIT, BFO, AGC, Speaker Out, Keypad, 200 Memories, Tone, RF Gain, Dimmer, Clock-Timer, RS-232.
Accs.: CFL-231 300 Hz Filter, CFL-232 500 Hz Filter, NVA-319 Speaker, CFL-233 1000 Hz Filter, CFL-218A 1800 Hz Filter, CMH-530 RTTY Demod., CGD-135 Hi Stab., CFL-243 BWC, CMF-78 ECSS.
New Price: $1200-1299 **Used Price:** $650-720 **Rating:** ★★★★★
Comments: The **NRD-535D** ($1700 new, $850 used) deluxe model includes the: CFL-243, CMF-78 and CFL-233. Model **NRD-535V** includes the CMF-78.

JRC *Japan Radio Co., Ltd.*
NRD-545

General Coverage Communications Receiver

Made In:	Japan 1998-1999	**Voltages:**	100/120/220/240 VAC
Coverage:	100-30000 kHz	**Readout:**	Digital Fluor. 0.01
Modes:	AM/SSB/CW/RTTY/FAX/FM	**Selectivity:**	See Comments
Circuit:	Triple Conversion	**Physical:**	13x5.2x11.3" 17 Lbs.

Features: ¼" Head. Jack, S-Indicator, Attenuator, NB, Mute Terminals, Line Out Jack, ANL, BFO, AGC, PBT, Speaker Output, Keypad, 1000 Memories, Tone, RF Gain, Dimmer, RS-232, Scan, Sweep, Synchronous Detection, Clock Timer, Timer Contacts, Notch.
Accs.: CGD-197 High Stability, NVA-319 Speaker, 6ZCJD00350 Cable, CHE-199 VHF-UHF Converter 30-1849 MHz (less cellular in USA).
New Price: $1700-1900 **Used Price:** (Too new) **Rating:** ★★★★★
Comments: The NRD-545 utilizes DSP (Digital Signal Processing) at the IF level. Bandwidth may be adjusted continuously from 40 Hz to 10 kHz in 10 Hz steps. Tuning steps: 1, 10, 100 Hz, 1, 5, 6.25, 9, 10, 12.5, 20, 25, 30 and 100 kHz.

11 Kenwood

Kenwood U.S.A.
P.O. Box 22745
Long Beach, CA 90801

KENWOOD R-300

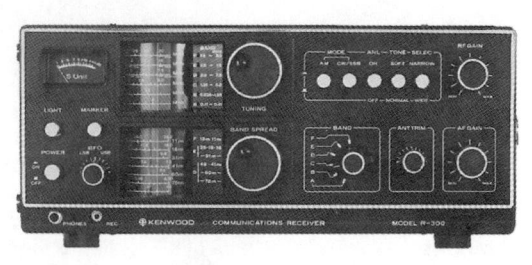

General Coverage Communications Receiver
Made In: Japan 1976-1979   **Voltages:** 117/220 VAC
Coverage: 170-30000 kHz   **Readout:** Analog
Modes: AM/SSB-CW   **Selectivity:** 5/2.5 kHz
Circuit: Double Conversion   **Physical:** 14.3x6.4x12.8" 17 Lbs.
Features: ¼" Head. Jack, S-Meter, Bandspread, ANL, Antenna Trimmer, Calibrator, BFO, Dial Lamp, Tone Switch, RF Gain, External Speaker Jack, Record Jack.
New Price: $239   **Used Price:** $80-95   **Rating:** ★★
Comments: Also operates from 12 VDC or eight D cells. Scarce.

KENWOOD R-600

General Coverage Communications Receiver
Made In: Japan 1982-1985   **Voltages:** 100/120/220/240 VAC
Coverage: 150-30000 kHz   **Readout:** Digital Phos. 1.
Modes: AM/USB/LSB-CW   **Selectivity:** 6/2.7 kHz
Circuit: Triple Conversion   **Physical:** 12.8x4.5x7.8" 10 Lbs.
Features: ¼" Head. Jack, S-Meter, Attenuator, Tone, NB, Mute Terminals, Record Jack, Carry Handle.
Accs.: DCK-1 12 VDC Kit.
New Price: $330-399   **Used Price:** $180-240   **Rating:** ★★★
Comments: An economy communications receiver aimed at the beginner. Not especially stable.

KENWOOD
R-1000

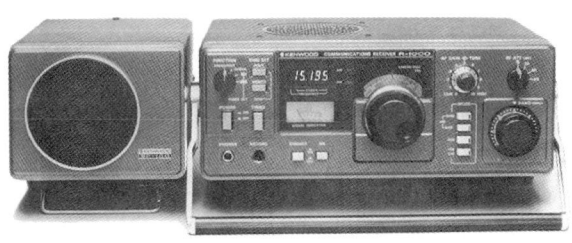

General Coverage Communications Receiver
Made In: Japan 1979-1985 **Voltages:** 100/120/220/240 VAC
Coverage: 200-30000 kHz **Readout:** Digital Phos. 1.
Modes: AM/USB/LSB/CW **Selectivity:** 12/6/2.7 kHz
Circuit: Double Conversion **Physical:** 12.8x4.5x8.6" 12 Lbs.
Features: ¼" Head. Jack, S-Meter, Attenuator, Tone, NB, Mute Terminals, Record Jack, Recorder Activation Jack, 12 Hour Clock-Timer, Tilt-Carry Handle, Dimmer.
Accs.: DCK-1 12 VDC Kit, SP-100 Speaker (shown).
New Price: $430-499   **Used Price:** $270-280   **Rating:** ★★★★★
Comments: A solid performer and very stable.

KENWOOD
R-2000

General Coverage Communications Receiver
Made In: Japan 1983-1992 **Voltages:** 100/120/220/240 VAC
Coverage: 150-30000 kHz **Readout:** Digital Phos. 0.1
Modes: AM/FM/USB/LSB/CW **Selectivity:** 6/2.7/_ kHz
Circuit: Triple Conversion **Physical:** 14.8x4.5x8.3" 12 Lbs.
Features: ¼" Head. Jack, S-Meter, Attenuator, Tone, NB, Mute Terminals, Record Jack, Dual Clock-Timer, Tilt Bar, Dimmer, 10 Memories, Squelch, Record Activation, Dial Lock, Scan, Sweep, Beep Level Adjust, Carry Handle, Three Tuning Rates.
Accs.: SP-100 Speaker, VC-10 Internal VHF Converter (118-174 MHz), YG-455C 500 Hz Filter, DCK-1 12 VDC Kit.
New Price: $499-699   **Used Price:** $370-420   **Rating:** ★★★★
Comments: A good value. The keys marked Ø through 9 are strictly for memory recall. They are not for frequency input.

KENWOOD
R-5000

General Coverage Communications Receiver

Made In:	Japan 1987-1996	**Voltages:**	120 VAC
Coverage:	100-30000 kHz	**Readout:**	Digital Phos. 0.01
Modes:	AM/FM/USB/LSB/CW	**Selectivity:**	6/2/_/_ kHz
Circuit:	Triple Conversion	**Physical:**	10.6x3.8x10.6" 12 Lbs.

Features: ¼" Head. Jack, S-Meter, Attenuator, Tone Control, Dual NB, Record Jack, Dual Clock-Timer, Tilt Bar, Dimmer, Squelch, 100 Memories, Record Activation, Dial Lock, Scan, Sweep, AGC, Mute Terminals, Carry Handle, Notch, IF Shift.

Accs.: DCK-2 12 VDC Kit, SP-430 Speaker, VC-20 VHF Converter, YG-88C 500 Hz Filter, YK-88CN 270 Hz CW Filter, YK-88A1 6 kHz AM Filter, YK-88SN 1.8 kHz SSB Filter, SP-430 Speaker, VS-1 Voice Synthesizer, IF-232/IC-10 Interface.

New Price: $799-1019   **Used Price:** $550-590   **Rating:** ★★★★★

Comments: The 2x5 keypad format takes some time to get used to. The keyboard tends to develop "key bounce" after several years of use, but otherwise a very reliable set. The optional VC-20 covers 108-174 MHz.

12 Lowe

Lowe Electronics Ltd.
Chesterfield Road,
Matlock, Derbyshire
England DE4 5LE

HF-125

General Coverage Communications Receiver
Made In:	England 1987-1988	**Voltages:**	12 VDC or 8 x AA cells
Coverage:	30-30000 kHz	**Readout:**	Digital LED 1.
Modes:	CW/AM/LSB/USB	**Selectivity:**	10/7/4/2.5 kHz
Circuit:	Double Conversion	**Physical:**	10x4x8.3" 4.2 Lbs.

Features: ¼" Head. Jack, 30 Memories, Attenuator, Record Jack, 400 Hz Audio CW Filter, Tone Control, Speaker Jack, Tilt Stand.
Accs.: D-125 Sync. Det./FM, K-125 Keypad, P-125 Portability Option
New Price: $599   **Used Price:** $370-380   **Rating:** ★★★
Comments: A nice clean receiver. Usually supplied with an AC adapter.

HF-150

General Coverage Communications Receiver
Made In:	England 1992-1999	**Voltages:**	12 VDC or 8 x AA cells
Coverage:	30-30000 kHz	**Readout:**	Digital LED 1.
Modes:	AM/LSB/USB	**Selectivity:**	7/2.5 kHz
Circuit:	Double Conversion	**Physical:**	7.2x3.2x6.5" 3 Lbs.

Features: ¼" Head. Jack, 60 Memories, Record Jack, Whip Preamp, Synchronous Detection, Two Tuning Rates, Attenuator.
Accs.: AP-150 Audio Processor, MB-150 Mobile Bracket, C-150 Case, RK-150 Rack Shelf, IF-150 PC Interface, PR-150 Preselector, KPAD-1 Keypad, KPAD-2 Keypad, AK-150 Accessory Kit
New Price: $499-699   **Used Price:** $340-370   **Rating:** ★★★★
Comments: A compact, robustly built receiver. The **HF-150E** *Europa* version (1998) includes a backlit LCD, a black case and other refinements (new $700). Model **HF-150M** is a marine version (new $650). All supplied with an AC adapter.

HF-225

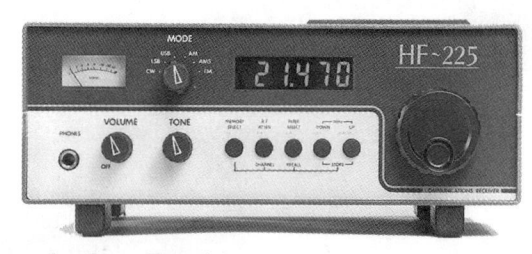

General Coverage Communications Receiver
Made In:	England 1990-1998	**Voltages:**	12 VDC
Coverage:	30-30000 kHz	**Readout:**	Digital LCD 1.
Modes:	CW/AM/LSB/USB	**Selectivity:**	10/7/4/2.5 kHz
Circuit:	Double Conversion	**Physical:**	10x4x8.3" 4.2 Lbs.

Features: ¼" Head. Jack, 30 Memories, Attenuator, Record Jack, Tilt Bar, 200 Hz Audio CW Filter, Backlit LCD, Speed Sensitive Tuning, Dual VFOs.
Accs.: D-225 Sync. Det & FM, KPAD-1 Wired Keypad, S-225 Speaker, C-225 Carry Case, B-225 NiCad Pack
New Price: $749-799   **Used Price:** $470-480   **Rating:** ★★★★
Comments: A nice clean receiver. Model **HF-225E** *Europa* has improved filters (new $1100, used $650). Usually supplied with an AC adapter.

HF-250

General Coverage Communications Receiver
Made In:	England 1995-1998	**Voltages:**	10-15 VDC
Coverage:	30-30000 kHz	**Readout:**	Digital LCD 0.1
Modes:	CW/AM/LSB/USB/FM	**Selectivity:**	10/7/4/2.2 kHz
Circuit:	Double Conversion	**Physical:**	11x4.13x5.2" 6 Lbs.

Features: ¼" Head. Jack, 255 Memories, Attenuator, Record Jack, 200 Hz Audio CW Filter, Dual VFOs, Tilt Feet, RS232 Port, FM Squelch, Backlit Display, Mute Line, Tone Control, Recorder Activation, Dual Clock Timer, 1 MHz Up-Down Tuning.
Accs.: DU-250 Sync. Det & FM, RC-250 Infrared Remote, WA-250 Telescopic Whip & Preamp
New Price: $1150-1200 **Used Price:** $580-650   **Rating:** ★★★
Comments: The model **HF-250E** *Europa* ($1300) has improved filters and chokes. Both models include an AC power supply.

13 Magnavox

Philips-Magnavox
1111 Northshore Dr.
Knoxville, TN 37919

MAGNAVOX D2935

General Coverage Portable Communications Receiver
Made In: Hong Kong 1986-1989   **Voltages:** 120/240 VAC
Coverage: 146-30000 kHz + FM   **Readout:** Digital LCD 1.
Modes: AM/SSB   **Selectivity:** 7 kHz
Circuit: Double Conversion   **Physical:** 12.5x7x3"
Features: ¼" Head. Jack, S Indicator, Tone, Keypad, Clock, RF Gain, 9 Memories, Line Out Jack, Two Speed Manual Tuning, BFO, Dial Lamp, Dial Lamp Switch, DX-Local Switch, Carry Strap, Antenna Jack, Tilt Stand, Speaker Output Jack.
New Price: $180-290   **Used Price:** $85-115   **Rating:** ★★★★
Comments: Also operates from six D cells and three AA cells or 12 VDC. The front panel utilizes membrane-type keys and keypad.

MAGNAVOX D2999

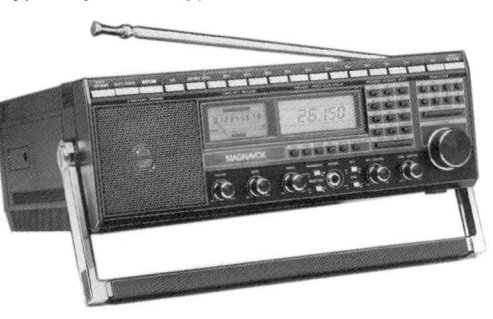

General Coverage Portable Communications Receiver
Made In: Hong Kong 1986-1989   **Voltages:** 110/127/220/240 VAC
Coverage: 150-26100 kHz + FM   **Readout:** Digital LCD 1.
Modes: AM/SSB   **Selectivity:** 6/3 kHz
Circuit: Double Conversion   **Physical:** 12.5x4.25x9.8" 11 Lbs.
Features: ¼" Head. Jack, S/Battery-Meter, Treble, Bass, Keypad, RF Gain, 16 Memories, Wide-Narrow, Sleep, Line Out Jack, Clock 12/24, 9/10 kHz MW Step, DX-Local Switch, Carry Handle, Woofer 7", BFO, AGC, Antenna Jack, Speaker Output, Two Tuning Rates.
New Price: $300-350   **Used Price:** $160-220   **Rating:** ★★★★
Comments: Also operates from six D cells and three AA cells or 12 VDC. The D2999 has uncommonly full audio as a result of the internal 7" and 3" speakers.

14 McKay Dymek

McKay Dymek Inc.
222 Kelso Dr.
North Bend, OR 97459

McKay Dymek
DR22C-6

General Coverage Communications Receiver
Made In: United States 1977-1980 **Voltages:** 110-120/220-240 VAC
Coverage: 50-29700 kHz **Readout:** Digital LED 5.
Modes: AM/LSB/USB **Selectivity:** 8/4 kHz
Circuit: Triple Conversion **Physical:** 17.5x5.1x15" 20 Lbs.
Features: ¼" Head. Jack, S-Meter, 5000 Hz Fixed Notch, NL, IF Output Jack, Mute Line, Fine Tuning.
Accs.: DS111 Speaker, 19RM22 Rack Kit, DP40 RF Preselector.
New Price: $995-1250 **Used Price:** $260-310 **Rating:** ★★★
Comments: One knob is featured for each digit of display except last digit which displays either 0 or 5 kHz. The Fine Tuning knob on this digit is for tuning stations not on an even 5 kHz channel. This series of receivers has a distinct "home stereo" look. Earlier model **DR22** has no NL. Both are very scarce.

McKay Dymek
DR33C-6

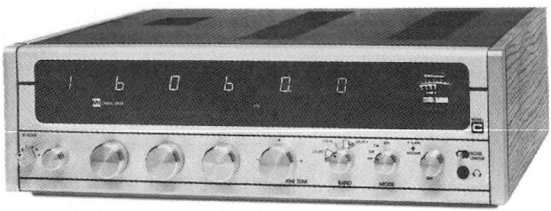

General Coverage Communications Receiver
Made In: United States 1975-1987 **Voltages:** 110-120/220-240 VAC
Coverage: 50-29700 kHz **Readout:** Digital LED 0.1
Modes: AM/LSB/USB/CW **Selectivity:** 8/4/_ kHz
Circuit: Triple Conversion **Physical:** 17.5x5.1x15" 16 Lbs.
Features: ¼" Head. Jack, S-Meter, 5000 Fixed Notch, NL, IF Output Jack, Four Tuning Steps, Preamp, Attenuator, Fine Tuning, Mute Line, Line Output Jack.
Accs.: DS111 Speaker, 19RM22 Rack Kit, DP40 RF Preselector, 375 Hz Filter, 1200 Hz Filter.
New Price: $1000-1995 **Used Price:** $400-560 **Rating:** ★★★
Comments: Very scarce.

ᴹᴷ McKay Dymek
DR44

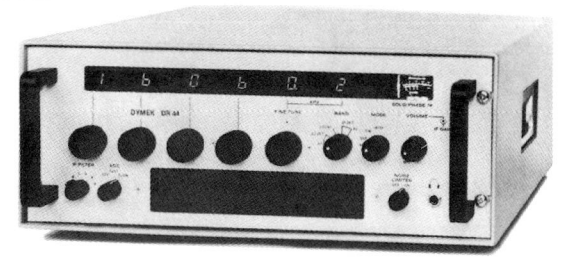

General Coverage Communications Receiver
Made In: United States 1975-1987 **Voltages:** 110-120/220-240 VAC
Coverage: 50-29700 kHz **Readout:** Digital LED 0.1
Modes: AM/SSB/CW/RTTY **Selectivity:** 8/4/2.5/_ kHz
Circuit: Triple Conversion **Physical:** 19x7x15" 16 Lbs.
Features: ¼" Head. Jack, S-Meter, 5000 Hz Fixed Notch, NL, Mute Line, IF Gain, IF Out Jack, Rack Handles, Fine Tuning.
Accs.: DS111 Speaker, DP4044 RF Preselector, 375 Hz Filter, 1200 Hz Filter.
New Price: $1600-1900 **Used Price:** $410-490 **Rating:** ★★★
Comments: Very scarce. The model **DR44-6** is similar, but with an additional knob for adjusting the AGC ($450-500 used).

ᴹᴷ McKay Dymek
DR101-6

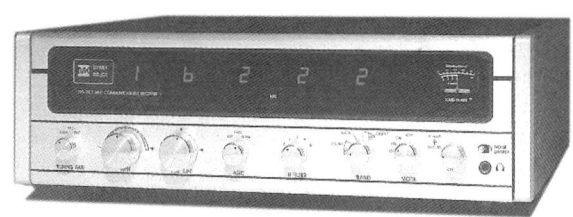

General Coverage Communications Receiver
Made In: United States 1979-1983 **Voltages:** 110-120/220-240 VAC
Coverage: 50-29700 kHz **Readout:** Digital LED 1
Modes: AM/SSB/CW/RTTY **Selectivity:** 8/4/2.5/_ kHz
Circuit: Triple Conversion **Physical:** 17.5x5.1x15" 16 Lbs.
Features: ¼" Head. Jack, S-Meter, 5000 Hz Fixed Notch, NL, IF Out Jack, IF Gain, Fine Tuning, Variable Rate Incremental Tuning, AGC, Mute Line, Three Tuning Rates.
Accs.: DS111 Speaker, DP40 RF Preselector, Rack Mounting Kit, 375 Hz Filter, 1200 Hz Filter.
New Price: $1150-1850 **Used Price:** $420-600 **Rating:** ★★★★
Comments: Allows manual variable rate sleigh tuning in 100 Hz steps. Scarce.

15 Panasonic

Panasonic
1 Panasonic Way
Secaucus, NJ 07094

Panasonic
RF-799

Portable Shortwave Broadcast Receiver
Made In: Japan 1982-1985 **Voltages:** 110~127/220~240 VAC
Coverage: See Comments **Readout:** Digital LCD 5.
Modes: AM **Selectivity:** 4 kHz
Circuit: Single Conversion **Physical:** 11.4x6.6x2.6" 4 Lbs.
Features: Mini Head. Jack, 10 Memories, Carry Handle, Lock, Keypad, Up-Down Tuning, Clock, LED Tune Indicator, Record Jack, Tone, Dial Lamp, Dial Lamp Switch, External Antenna Input.
New Price: $240-300 **Used Price:** $70-80 **Rating:** ★★
Comments: Ranges: .153-.281, .53-1.61, 2.3-2.935, 2.94-3.575, 3.58-4.215, 4.54-5.175, 5.82-6.455, 7.1-7.735. 9.5-10.135, 11.58-12.215, 15.1-15.735, 17.5-18.135, 21.34-21.975 and 25.5-26.135 MHz plus FM. Operates from four C and two AA cells. The **RF-799LBS** and **RF-799LBE** are European versions.

Panasonic
RF-2600

General Coverage Portable Communications Receiver
Made In: Japan 1979-1981 **Voltages:** 110-125/220-240 VAC
Coverage: See Comments **Readout:** Digital Fluor. 1.
Modes: AM/SSB-CW **Selectivity:** 16/4 kHz
Circuit: Double Conversion **Physical:** 13.5x9.3x4.6" 7 Lbs.
Features: Mini Head. Jack, S-Meter, BFO, MPX Output, FM AFC, Bass, Treble, Shoulder Strap, Record Jack, Two Tuning Rates, Dial Lamp, Dial Lamp Switch, RF Gain, Calibrator, Phono Input, External Antenna Input.
New Price: $200-250 **Used Price:** $90-110 **Rating:** ★★★
Comments: Ranges: .525-16.1, 3.9-10, 10-16, 16-22 and 22-28 MHz plus FM. Also operates from six D cells. Sold in Europe as model **DR26**.

Panasonic
RF-2800

General Coverage Portable Communications Receiver
Made In: Japan 1978-1979 **Voltages:** 120 VAC 60 Hz
Coverage: See Comments **Readout:** Digital LED 1.
Modes: AM/SSB/CW **Selectivity:** 5/3.4 kHz
Circuit: Double Conversion **Physical:** 15x9.8x4.75" 9 Lbs.
Features: Mini Head. Jack, S-Meter, BFO, MPX Output, FM AFC, Bass, Treble, Shoulder Strap, Record Jack, Two Tuning Rates, Dial Lamp, Dial Lamp Switch, RF Gain, External Antenna Input.
New Price: $200-250   **Used Price:** $80-95   **Rating:** ★★★
Comments: Ranges: .525-1.61, 3.2-8, 8-16 and 16-30 MHz plus FM. Operates from six D cells. Not as strong a performer as its analog predecessor, the RF-2200. Model **DR28** is the European version. The later **RF-2900** features a blue fluorescent digital display (European version is **DR29**).

Panasonic
RF-3100

General Coverage Portable Communications Receiver
Made In: Japan 1982-1985 **Voltages:** 120 VAC
Coverage: 525-30000 kHz +FM **Readout:** Digital Fluor. 1.
Modes: AM/SSB **Selectivity:** 7/3 kHz
Circuit: Double Conversion **Physical:** 14.6x4.8x9.5" 8 Lbs.
Features: ¼" Head. Jack, S-Meter, S-Meter Lamp, S-Meter Lamp Switch, RF Gain, BFO, ANL, Tilt Bar, Bass, Treble, Record Jack, Speaker Jack, Shoulder Strap, External Antenna Jack.
New Price: $289-319   **Used Price:** $150-160   **Rating:** ★★★
Comments: Pleasant audio and a good value. Also operates from eight D cells. Sold outside the U.S. as models **DR31** or **RF-B30**.

Panasonic
RF-4800

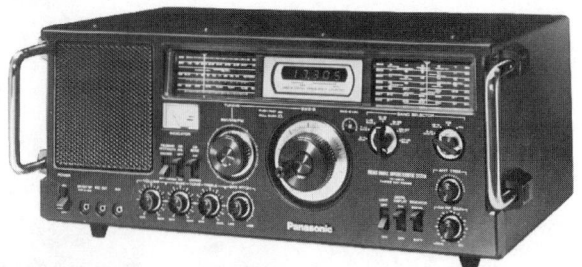

General Coverage Communications Receiver
Made In: Japan 1976-1978 **Voltages:** 120 VAC or 12 VDC
Coverage: 525-31000 kHz +FM **Readout:** Digital LED 1.
Modes: AM/SSB/CW **Selectivity:** 5/3.4 kHz
Circuit: Double Conversion **Physical:** 19x7.8x14" 20 Lbs.
Features: ¼" Head. Jack, S/Battery-Meter, RF Gain, BFO, ANL, Light, Bass, Treble, Speaker Jack, Two Tuning Speeds, Handles, Antenna Trimmer, Spinner Knob, Record Jack, Aux. Input Jack.
New Price: $450   **Used Price:** $150-160   **Rating:** ★
Comments: Also operates from eight D cells. More properly classified as an analog receiver with digital counter than a true digital receiver. The band switch tends to wear and replacement is difficult. Not stable. **DR48** outside U.S.A.

Panasonic
RF-4900

General Coverage Communications Receiver
Made In: Japan 1979-1983 **Voltages:** 120 VAC or 12 VDC
Coverage: 525-31000 kHz +FM **Readout:** Digital Fluor. 1.
Modes: AM/SSB/CW **Selectivity:** 5/3.4 kHz
Circuit: Double Conversion **Physical:** 19x7.8x14" 20 Lbs.
Features: ¼" Head. Jack, S/Battery-Meter, RF Gain, BFO, ANL, Bass, Treble, Speaker Jack, Two Tuning Speeds, Handles, Antenna Trimmer, Spinner Knob, Light, Record Jack, Auxiliary Audio Input Jack.
New Price: $399-479   **Used Price:** $170-190   **Rating:** ★★
Comments: Also operates from eight D cells. The RF-4900 is similar to the earlier RF-4800, but with a green fluorescent digital display rather than a red LED digital display. Sold outside the United States as the model **DR49**.

Panasonic
RF-6300

General Coverage Portable Communications Receiver
Made In: Japan 1982-1983 **Voltages:** See Comments
Coverage: 150-30000 kHz +FM **Readout:** Digital Phs. 1.
Modes: AM/SSB-CW **Selectivity:** 5/3.4 kHz
Circuit: Double Conversion **Physical:** 17.2x11.2x5.3" 12 Lbs.
Features: ¼" Head. Jack, S-Battery Meter, Clock, 12 Memories, Lock, Bass, Treble, Carry Handle, Record Jack, Two Tuning Rates, LCD Clock, Dial Lamp, Dial Lamp Switch, RF Gain, Ext. Ant. J.
New Price: $500-700 **Used Price:** $170-210 **Rating:** ★★★
Comments: Operates from: 100~110/115~127/200~220/230~250 VAC 50/60 Hz or six D and four AA Cells or 9 VDC. Scarce. Model **RF-6300L** includes .15-.41 MHz longwave coverage. Also sold outside the U.S. as model **DR Q 63**.

Panasonic
RF-9000

General Coverage Portable Communications Receiver
Made In: Japan 1982-1985 **Voltages:** See Comments
Coverage: 150-30000 kHz +FM **Readout:** Digital LCD 1.
Modes: AM/SSB-CW **Selectivity:** 4.8/3.2/2.4 kHz
Circuit: Double Conversion **Physical:** 20.6x14.3x8.2" 45 Lbs.
Features: ¼" Head. Jack, S-Tune Meter, Clock, 15 Memories, Lock, ANL, Bass, Treble, Carry Handle, Record Jack, Two Tuning Rates, Clock, Dial Lamp, Dial Lamp Switch, RF Gain, Keypad, Scanning, Aux. Audio Input Jack, 12/24 Hour Clock, Four Event Timer, External Antenna Input, Loudness, Two Speakers 7x5" & 2.6".
New Price: $2900-3800 **Used Price:** $780-900 **Rating:** ★
Comments: Operates from: 100~110/115~127/200~220/230~250 VAC or 8 D cells or 12 VDC. Extremely scarce due to low production and high collector demand.

Panasonic
RF-B40

General Coverage Portable Broadcast Receiver
Made In: Japan 1989-1990 **Voltages:** 6 VDC
Coverage: 146-29995 kHz +FM **Readout:** Digital LCD 5.
Modes: AM **Selectivity:** One Position
Circuit: Double Conversion **Physical:** 7.4x4.3x1.3" 1.2 Lbs.
Features: Mini Head. Jack, S Indicator, 27 Memories, 9/10 kHz MW Step, Tilt Stand, Lock, Sleep, Clock, DX-Local Switch, Tone Switch, Keypad, Up-Down Tuning, Two Tuning Speeds, Wrist Strap, External Antenna Terminals.
Accs.: RP-63 AC Adapter, RP-993 Car Power Adapter
New Price: $200-230 **Used Price:** $65-75 **Rating:** ★★★
Comments: Also operates from four AA cells. The 27 memories are allocated as follows: 9 for FM, 9 for LW/MW and 9 for SW. Supplied with earphone and carrying strap.

Panasonic
RF-B45

General Coverage Portable Communications Receiver
Made In: Japan 1991-1997 **Voltages:** 6 VDC
Coverage: 155-30000 kHz +FM **Readout:** Digital LCD 1.
Modes: AM/SSB **Selectivity:** 6 kHz
Circuit: Double Conversion **Physical:** 8.2x4.4x1.6" 1.4 Lbs.
Features: Mini Head. Jack, S Indicator, 18 Memories, 9/10 kHz MW Step, Tilt Stand, DX-Local Switch, Lock, Sleep, Dual Clock, Keypad, Up-Down Tuning, Wrist Strap, Tone Switch, Two Tuning Speeds, Fine Tuning, External Antenna Jack.
Accs.: RP-65 AC Adapter
New Price: $150-200 **Used Price:** $80-85 **Rating:** ★★★
Comments: Also operates from four AA cells. The 18 memories are allocated as follows: 9 for FM and 9 for LW/MW/SW. Supplied with earphone and carrying pouch.

Panasonic
RF-B60

General Coverage Portable Broadcast Receiver
Made In: Japan 1989-1991   **Voltages:** 6 VDC
Coverage: 155-30000 kHz +FM   **Readout:** Digital LCD 1.
Modes: AM   **Selectivity:** 6 kHz
Circuit: Double Conversion   **Physical:** 7.8x4.7x1.4" 1.4 Lbs.
Features: Mini Head. Jack, S Indicator, 36 Memories, 9/10 kHz MW Step, Tilt Stand, DX-Local Switch, Dual Clock Timer, Wrist Strap, Keypad, Up-Down Tuning, Tone Switch, Manual Tuning, Two Tuning Speeds, Sleep, Dial-Keypad Lock.
Accs.: RP-63 AC Adapter, RP-65 AC Adapter, RP-993 Auto Converter.
New Price: $240-320   **Used Price:** $85-90   **Rating:** ★★★
Comments: Also operates from six AA cells. The 36 memories are allocated as follows: 9 for FM, 9 for LW, 9 for MW and 9 for SW. Supplied with earphone and carrying pouch. Note that single sideband reception is not supported.

Panasonic
RF-B65

General Coverage Portable Communications Receiver
Made In: Japan 1989-1992   **Voltages:** 6 VDC
Coverage: 155-30000 kHz +FM   **Readout:** Digital LCD 1.
Modes: AM/SSB   **Selectivity:** 6 kHz
Circuit: Double Conversion   **Physical:** 7.8x4.7x1.4" 1.4 Lbs.
Features: Mini Head. Jack, S Indicator, 36 Memories, 9/10 kHz MW Step, Tilt Stand, DX-Local Switch, Dial Lock, Sleep, Dual Clock Timer, Up-Down Tuning, Manual Tuning, Wrist Strap, Tone Switch, Keypad, Keypad Lock, Two Tuning Speeds, Fine Tuning.
Accs.: RP-63 AC Adapter, RP-65 AC Adapter, RP-993 Auto Converter.
New Price: $280-350   **Used Price:** $90-100   **Rating:** ★★★★
Comments: Adds fine tuning and SSB capability to the RF-B60. Also operates from six AA cells. The 36 memories are allocated as follows: 9 for FM, 9 for LW, 9 for MW and 9 for SW. Supplied with earphone and carrying pouch.

Panasonic

RF-B300

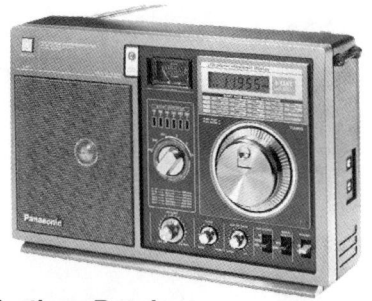

General Coverage Portable Communications Receiver

Made In:	Japan 1983-1985	**Voltages:**	110~127/220~240 VAC
Coverage:	520-30000 kHz +FM	**Readout:**	Digital LCD 5.
Modes:	AM/USB-CW/LSB-CW	**Selectivity:**	7/3 kHz
Circuit:	Double Conversion	**Physical:**	13.7x8.5x4.25" 5 Lbs.

Features: Mini Head. Jack, S-Meter, Antenna Terminals, Wide-Narrow, Dial Lamp, Dial Lamp Switch, RF Gain, Tone Control, Record Jack, Slow/Fast Tuning, Dial Lock, Carry Strap.

New Price: $220-250   **Used Price:** $90-110   **Rating:** ★★★

Comments: Also operates from six C cells. Sold as model **DR-B300** in some areas of the world.

Panasonic

RF-B600

General Coverage Portable Communications Receiver

Made In:	Japan 1984-1989	**Voltages:**	120/240 VAC 12 VDC
Coverage:	150-29999 kHz +FM	**Readout:**	Digital Fluor. 1.
Modes:	AM/LSB/USB	**Selectivity:**	7/3.5 kHz
Circuit:	Double Conversion	**Physical:**	14.8x4.8x11.5" 10 Lbs.

Features: ¼" Head. Jack, S-Battery Meter, Meter Lamp Switch, Lock, Carry Handle, 9 Memories, Tilt Stand, Mute Jack, DC Input Jack, Memory Scan, Keypad, Sweep (Zone Scan), Record Jack, Two Speed Manual Tuning, Digital Display On-Off Switch, External Antenna Jack.

Accs.: RP-952 Car PS Adapter

New Price: $440-650   **Used Price:** $250-280   **Rating:** ★★★

Comments: Requires three AA cells and may also operate from eight D cells. Also sold as model **DR-B600** outside the U.S.

16 Realistic

Radio Shack
P.O. Box 2625
Forth Worth, TX 76113

REALISTIC

DX-100

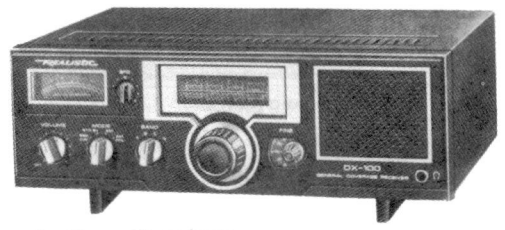

General Coverage Communications Receiver
Made In: Taiwan 1981-1984 **Voltages:** 117 VAC
Coverage: 550-30000 kHz **Readout:** Analog
Modes: AM/SSB-CW **Selectivity:** One Position
Circuit: Single Conversion **Physical:** 12x5.7x8"
Features: ¼" Head. Jack, S-Meter, ANL, Dial Lamp, Standby, Fine Tuning.
New Price: $100   **Used Price:** $30-35   **Rating:** ★
Comments: The DX-100 also operates from 12 VDC. A very basic receiver. Please note that Realistic is a registered trademark of Tandy Corporation.

REALISTIC

DX-120

General Coverage Communications Receiver
Made In: Japan 1970-1971 **Voltages:** 117 VAC 12 VDC
Coverage: 535-30000 kHz **Readout:** Analog
Modes: AM/SSB-CW **Selectivity:** One Position
Circuit: Single Conversion **Physical:** 12.5x5.25x7.8" 10 Lbs.
Features: ¼" Head. Jack, S-Meter, ANL, Dial Lamp, RF Gain, Bandspread 0-100, Dial Lamp, AVC.
Accs.: SP-150 Speaker, DC Pack (8xD)
New Price: $70   **Used Price:** $30-40   **Rating:** ★★
Comments: The full model name is the Realistic DX-120 Star Patrol.

REALISTIC
DX-150

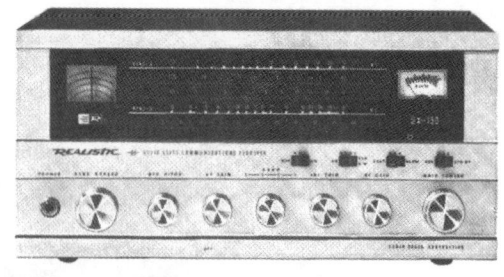

General Coverage Communications Receiver
Made In: Japan 1967-1969 **Voltages:** 120 VAC 12 VDC
Coverage: 535-30000 kHz **Readout:** Analog
Modes: AM/SSB-CW **Selectivity:** 4.5 kHz
Circuit: Single Conversion **Physical:** 14.2x6.5x9.25" 14 Lbs.
Features: ¼" Head. Jack, S-Meter, Mute Line, ANL, Dial Lamp, Standby, Bandspread, Antenna Trimmer, AVC.
Accs.: SP-150 Speaker, DC Pack (8xD)
New Price: $120 **Used Price:** $45-55 **Rating:** ★
Comments: Although this radio presented a good value for its era, its lack of selectivity and frequency accuracy make it weak choice today. Ranges: .535-1.6, 1.55-4.5, 4.5-13 and 13-30 MHz.

REALISTIC
DX-150A

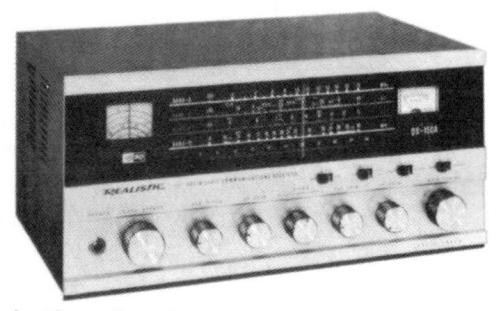

General Coverage Communications Receiver
Made In: Japan 1969-1972 **Voltages:** 120 VAC 12 VDC
Coverage: 535-30000 kHz **Readout:** Analog
Modes: AM/SSB-CW **Selectivity:** 4.5 kHz
Circuit: Single Conversion **Physical:** 14.2x6.5x9.25" 15 Lbs.
Features: ¼" Head. Jack, S-Meter, Mute Line, ANL, Dial Lamp, Standby, Bandspread, Antenna Trimmer, AVC.
Accs.: SP-150 Speaker, DC Pack (8xD)
New Price: $120 **Used Price:** $45-55 **Rating:** ★
Comments: Main tuning bands: .535-1.6, 1.55-4.5, 4.5-13 and 13-30 MHz. Bandspread bands: 3.5-4, 7-7.3, 14-14.35, 21-21.4, 28-29.5 MHz and CB channels 1-23. Similar to the DX-150, but with a FET front-end.

REALISTIC
DX-150B

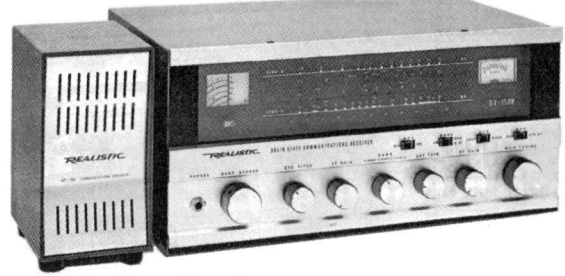

General Coverage Communications Receiver

Made In:	Japan 1972-1974	**Voltages:**	120 VAC 12 VDC
Coverage:	535-30000 kHz	**Readout:**	Analog
Modes:	AM/SSB-CW	**Selectivity:**	4.5 kHz
Circuit:	Single Conversion	**Physical:**	14.2x6.5x9.25" 15 Lbs.
Features:	¼" Head. Jack, S-Meter, Mute Line, ANL, Dial Lamp, Standby, Bandspread, Antenna Trimmer, AVC.		
Accs.:	DC Pack (8xD)		

New Price: $120-140   **Used Price:** $60-65   **Rating:** ★★

Comments: Ranges: .535-1.6, 1.55-4.5, 4.5-13 and 13-30 MHz. Bandspread bands: 3.3-4.1, 4.85-5.1, 5.75-6.25, 6.6-7.4, 8.4-10, 9.5-12.5, 13.7-14.4, 14.7-15.5, 16.8-18.1, 19.8-22 and 25.5-30 MHz. Supplied with matching external speaker (shown).

REALISTIC
DX-160

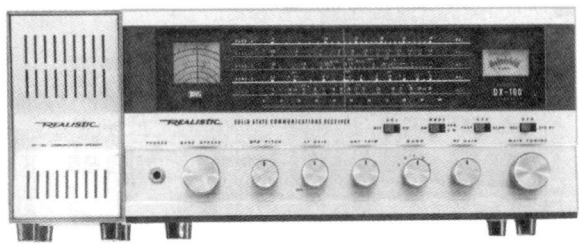

General Coverage Communications Receiver

Made In:	Japan 1975-1980	**Voltages:**	120 VAC 12 VDC
Coverage:	150-30000 kHz	**Readout:**	Analog
Modes:	AM/SSB-CW	**Selectivity:**	4 kHz
Circuit:	Single Conversion	**Physical:**	14.5x6.5x9.25" 15 Lbs.
Features:	¼" Head. Jack, S-Meter, Mute Line, ANL, Dial Lamp, Standby, Bandspread, Antenna Trimmer, AVC.		
Accs.:	DC Pack (8xD)		

New Price: $160   **Used Price:** $70-75   **Rating:** ★★

Comments: Ranges: .15-.4, .535-1.6, 1.55-4.5, 4.5-13 and 13-30 MHz. (Coverage gap from .4-.535 MHz.) Bandspread bands: 3.5-4, 7-7.3, 14-14.35, 21-21.4 and 28-29.7 MHz plus CB channels 1-23. Supplied with a matching external speaker (shown). Note the additional coverage of the longwave band.

Realistic
DX-200

General Coverage Communications Receiver

Made In:	Japan 1981-1983	**Voltages:**	120 VAC 12 VDC
Coverage:	150-30000 kHz[1]	**Readout:**	Analog
Modes:	AM/SSB-CW	**Selectivity:**	4 kHz
Circuit:	Single Conversion	**Physical:**	14.5x6.5x9.25"
Features:	¼" Head. Jack, S-Meter, Mute Line, ANL, Dial Lamp, Standby, Bandspread, Antenna Trimmer, Calibrator, AGC.		

New Price: $230   **Used Price:** $80-90   **Rating:** ★★

Comments: Ranges: .15-.4, .52-4.5, 4.5-13 and 13-30 MHz. [1]Coverage gap from .4-.52 MHz. Bandspread bands: 3.3-4.1, 4.85-5.1, 5.75-6.25, 6.6-7.4, 8.4-10, 9.5-12.5, 13.7-14.4, 14.7-15.5, 16.8-18.1, 19.8-22 and 25.5-30 MHz.

Realistic
DX-300

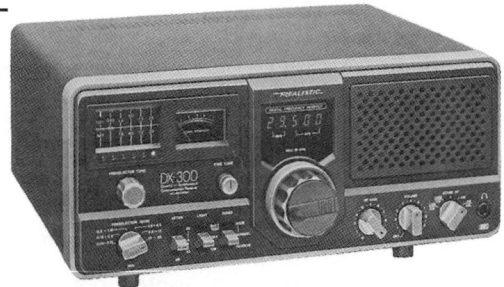

General Coverage Communications Receiver

Made In:	Japan 1979-1980	**Voltages:**	120 VAC 12 VDC
Coverage:	10-30000 kHz	**Readout:**	Digital LED 1.
Modes:	AM/LSB/USB-CW	**Selectivity:**	6 kHz
Circuit:	Triple Conversion	**Physical:**	14.2x6.5x9.25" 14 Lbs.
Features:	¼" Head. Jack, S-Meter, Mute Line, ANL, Dial Lamp, Standby, Fine Tuning, Audio Filter, Antenna Trimmer, AGC, RF Gain Preselector, Attenuator, Key Input Jack, Dial Lamp Switch.		

New Price: $380   **Used Price:** $120-140   **Rating:** ★

Comments: Can hold eight "C" cells for portable operation. This model lacked adequate stability and image rejection. Note that the switch for Wide/Normal/Narrow is merely an audio filter circuit and not an IF filter selection.

REALISTIC
DX-302

General Coverage Communications Receiver
Made In: Japan 1981-1982
Coverage: 10-30000 kHz
Modes: AM/LSB/USB-CW
Circuit: Triple Conversion
Voltages: 120 VAC 12 VDC
Readout: Digital LED 1.
Selectivity: 7/5 kHz
Physical: 14.2x6.5x9.25" 14 Lbs.
Features: ¼" Head. Jack, S-Meter, Mute Line, ANL, Dial Lamp, Standby, Fine Tuning, Key Input Jack, Attenuator, Antenna Trimmer, Preselector, Dial Lamp Switch, Record Jack, AGC.
New Price: $400   **Used Price:** $150-160   **Rating:** ★★
Comments: Can hold eight "C" cells for portable operation. This model lacked adequate stability and image rejection. Note that, unlike the DX-300, the switch for Wide/Narrow on the DX-302 is for a true IF filter selection.

REALISTIC
DX-394

General Coverage Communications Receiver
Made In: Japan 1995-1998
Coverage: 150-30000 kHz
Modes: AM/LSB/USB/CW
Circuit: Double Conversion
Voltages: 120 VAC 12 VDC
Readout: Digital LCD 0.1
Selectivity: 7.2/6/5.7 kHz
Physical: 9.125x3.5x7.8" 5 Lbs.
Features: Mini Head. Jack, S-Meter, VRIT Tuning, Dial Lamp, Keypad, NB, Fine Tuning, Clock, 5-Event Timer, Scan, Sweep, Standby, Lock, Record Jack, Sleep Timer, Four Tuning Steps.
New Price: $250-400   **Used Price:** $120-130   **Rating:** ★★★
Comments: The bandwidths are mode-dependent. The AC power supply is built in. A telescopic antenna is also featured.

17 Sangean

Sangean America
2651 Troy Ave.
South El Monte, CA 91733

SANGEAN
ATS-803A

General Coverage Portable Communications Receiver
Made In: Taiwan 1992-1995
Coverage: 150-30000 kHz + FM^S
Modes: AM/SSB
Circuit: Double Conversion
Voltages: 120 VAC
Readout: Digital LCD 1.
Selectivity: Two position
Physical: 11.5x6.3x2.4" 3.8 Lbs.
Features: Mini Head. Jack, S/Battery Indicator, Tone, Keypad, Clock-Timer, 9 Mems., Ext. Ant. Jack, Sleep, RF Gain, Lock, Ext. Ant., Handle, Record Output, Dial Lamp, Manual Tuning, Up-Dn Tuning, BFO.
New Price: $169-249   **Used Price:** $100-120   **Rating:** ★★★★★
Comments: Good single sideband reception for a radio in this price class. This popular model provides great value. Also 6 D and 2 AAs. Model **ATS-803** was earlier (1985). **Realistic DX-440** similar sans Record Output Jack and AC PS.

SANGEAN
ATS-808

General Coverage Portable Broadcast Receiver
Made In: Taiwan 1993-1998
Coverage: 150-30000 kHz + FM^S
Modes: AM
Circuit: Double Conversion
Voltages: 6 VDC
Readout: Digital LCD 1.
Selectivity: Two position
Physical: 7.75x4.5x1.5" 2 Lbs.
Features: Mini Head. Jack, S/Battery Indicator, Tone, Keypad, Sleep, Clock-Timer, 45 Memories, Manual Tuning, Up-Dn Tuning, Local-DX Switch, 9/10 kHz MW Step, Ext. Antenna Jack, Lock.
Accs.: ADP-808 AC Adapter
New Price: $119-249   **Used Price:** $60-70   **Rating:** ★★★★
Comments: Operates from six AA cells. The ATS-808 is a sensitive receiver with excellent FM and shortwave reception. It never gained wide popularity among DXers because of its lack of SSB reception. The **Realistic DX-380** is similar. Model **ATS-808A** (new $130-$200) is later production with 54 memories.

SANGEAN
ATS-818CS

General Coverage Portable Communications Receiver
Made In: Taiwan 1993-1999 **Voltages:** 120 VAC 6 VDC
Coverage: 150-30000 kHz + FM^S **Readout:** Digital LCD 1.
Modes: AM/SSB **Selectivity:** Two position
Circuit: Double Conversion **Physical:** 11.25x7x2.5" 5 Lbs.
Features: Mini Head. Jack, S/Batt. Indicator, Tone, Keypad, Clock-Timer, 45 Memories, Dial Lamp, Sleep, RF Gain, Lock, Carry Handle, BFO, Auto-Stop Mono Cassette Recorder, Mic, Up-Dn Tuning.
New Price: $199-299 **Used Price:** $90-100 **Rating:** ★★★★
Comments: Also operates from four C cells and three AA cells. The **Realistic DX-392** is similar. The model **ATS-818** (new $140-$200, used $80-$90) does not feature the built-in cassette recorder. The **Realistic DX-390** is similar.

SANGEAN
ATS-909

General Coverage Portable Communications Receiver
Made In: Taiwan 1997-1999 **Voltages:** 120 VAC 6 VDC
Coverage: 153-30000 kHz + FM^S **Readout:** Digital LCD 1.
Modes: AM/LSB/USB **Selectivity:** Two Position
Circuit: Double Conversion **Physical:** 8.5x5.5x1.5" 2 Lbs.
Features: Mini Head. Jack, S/Battery Indicator, Tone Switch, Keypad, 306 Alpha-Memories, Wide-Narrow, Sleep, RF Gain, Lock, RDS, Three Event Clock-Timer, Sweep, Record Output, Dial Lamp, Lamp Switch, 9/10 kHz MW Step, Up-Dn Tuning, Manual Tuning.
New Price: $249-299 **Used Price:** $140-150 **Rating:** ★★★★★
Comments: Operates from four AA cells. 260 of the 306 memories are available for shortwave. Supplied with carry case, windup antenna, stereo earphones and AC adapter. Tuning steps down to 40 Hz. Very good SSB performance for a portable. This feature-rich radio is Sangean's flagship model. The **Realistic DX-398** is similar.

18 Sony

Sony
1 Sony Dr.
Park Ridge, NJ 07656

SONY®
CRF-1

General Coverage Portable Communications Receiver
Made In: Japan 1981-1986   **Voltages:** 100/120/220/240 VAC
Coverage: 10-30000 kHz   **Readout:** Digital LED 0.1
Modes: AM/USB/LSB/CW   **Selectivity:** 10/4.4/2 kHz
Circuit: Double Conversion   **Physical:** 10.4x4x13.2" 15 Lbs.
Features: ¼" Head. Jack, S/Battery-Meter, Record Jack, Preselector, NB, Dial Lamp, Dial Lamp Switch, Carry Handle, Muting, RF Gain.
Accs.: DCC-9 Car Battery Cord
New Price: $1350-1795   **Used Price:** $500-650   **Rating:** ★★★
Comments: Also operates from 12 VDC or eight D and two AA cells. This early digital portable utilized 21 Integrated circuits, 21 FETs, 68 transistors and 99 diodes. This substantial model is very scarce and collectable.

SONY®
CRF-320A

General Coverage Portable Communications Receiver
Made In: Japan 1976-1980   **Voltages:** 110/120/220/240 VAC
Coverage: 150-29999 kHz +FM   **Readout:** Digital LED 1.
Modes: AM/USB/LSB/CW   **Selectivity:** 8/6 kHz
Circuit: Double Conversion   **Physical:** 17.8x12.2x8.2" 29 Lbs.
Features: ¼" Head. Jack, S/Battery Meter, Bass, Treble, RF Gain, NB, Dial Lamp, Clock Timer, Dial Adjust, Local-DX Switch, Antenna Trimmer, Carry Handle, Record Jack, Mute, Aux. Input Jack.
Accs.: DCC-9 Car Battery Cord
New Price: $1495   **Used Price:** $500-650   **Rating:** ★★★
Comments: Also operates from 12 VDC. May also operate from eight D cells. One additional D cell is also required for the clock. Scarce.

SONY®
CRF-330K

General Coverage Portable Communications Receiver
Made In: Japan 1978-1980 **Voltages:** 110/120/220/240 VAC
Coverage: 150-29999 kHz +FM **Readout:** Digital LED 1.
Modes: AM/USB/LSB/CW **Selectivity:** 8/6 kHz
Circuit: Double Conversion **Physical:** 17.8x12.2x8.2" 34 Lbs.
Features: ¼" Head. Jack, S/Battery Meter, Bass, Treble, RF Gain, Dial Lamp, Clock Timer, NB, Dial Adjust, Local-DX Switch, Carry Handle, Record Jack, Aux. Input Jack, Muting, Slide-Out Cassette Recorder, Antenna Trimmer, Built-in Mic.
Accs.: DCC-9 Car Battery Cord
New Price: $2495   **Used Price:** $580-700   **Rating:** ★★★
Comments: Also operates from 12 VDC or eight D cells. One D cell is also required for the clock. There is strong collector demand for this scarce model.

SONY®
CRF-V21

General Coverage Portable Communications Receiver
Made In: Japan 1989-1992 **Voltages:** 110/120/220/240 VAC
Coverage: 9-30000 kHz +FM **Readout:** Digital LCD 0.1
Modes: AM/USB/LSB-CW/FM/RTTY/FAX **Selectivity:** 6/3.5/2.7/14 kHz
Circuit: Double Conversion **Physical:** 16.3x11.3x6.8" 21 Lbs.
Features: Mini Ear Jack, S Meter, 350 Alpha Memories, Carry Handle, Active Antenna, Clock, 8 Event Timer, Scan, Sweep, FM AFC, Sync. Detect., Atten., AF Filter, Record Jack, Dial Lamp, Keypad, Spectrum Display, LCD Contrast Adjust, RTTY/FAX Decoder,
Accs.: AN-P1200 GOES Satellite Antenna
New Price: $3500-6500 **Used Price:** $650-820   **Rating:** ★★★
Comments: Also operates from 6 VDC NiCad pack. Two AA cells are also required for the CPU. This technology show piece model is best known for its built-in spectrum display (200 kHz or 5 MHz) and RTTY/FAX decoding. Direct GOES satellite reception with optional AN-P1200 antenna. A built-in thermal printer provides hardcopy of RTTY, FAX or GOES receptions. Very scarce.

SONY®
ICF-2001

General Coverage Portable Communications Receiver
Made In: Japan 1980-1983 **Voltages:** 117 VAC or 4.5 VDC
Coverage: 150-29999 kHz +FM **Readout:** Digital LCD 1.
Modes: AM/SSB-CW **Selectivity:** One Position
Circuit: Double Conversion **Physical:** 12.8x6.8x2.2" 4 Lbs.
Features: Mini Head. Jack, S-Indicator, Bass, Treble, 6 Memories, Sweep, BFO, Keypad, Clock-Timer, Tilt Stand, Up-Down Tuning, Sleep, Antenna Trimmer, Antenna Terminals, Record Jack, Dial Lamp, Dial Lamp Switch, Rec. Activation, Two Tuning Speeds, Local-DX.
Accs.: DCC-127A Car Battery Cord.
New Price: $170-350   **Used Price:** $80-85   **Rating:** ★★★
Comments: Also operates from three D and two AA cells. Supplied with carry strap and earphone. In the U.S.A. the AC-120 adapter was included. A bit of a battery pig.

SONY®
ICF-2002

General Coverage Portable Communications Receiver
Made In: Japan 1983-1986 **Voltages:** 6 VDC
Coverage: 150-30000 kHz +FM **Readout:** Digital LCD 5.
Modes: AM/SSB-CW **Selectivity:** ±5.6 kHz
Circuit: Double Conversion **Physical:** 7.25x4.75x1.25" 1.5 Lbs.
Features: Mini Head. Jack, Tone Switch, 10 Memories, Sweep,
  Fine Tuning, 12/24 Clock-Timer, Up-Down Tuning, Keypad,
  LED Tune Indicator, 9/10 kHz MW Step, Attenuator,
  Record Output Jack, DX-Local Switch, External Antenna Jack.
Accs.: AC-9W AC Adapter, DCC-127A Car Battery Cord.
New Price: $210-250   **Used Price:** $90-100   **Rating:** ★★★★★
Comments: Supplied with carry case, BP-25 external battery case (4xC cells) and earphone. Operates from six AA cells. Light gray case. Sold as the model **ICF-7600D** outside U.S. (A dark gray version of the ICF-7600D also exists).

SONY®
ICF-2003

General Coverage Portable Communications Receiver
Made In: Japan 1987-1991 **Voltages:** 6 VDC
Coverage: 153-30000 kHz +FM **Readout:** Digital LCD 5.
Modes: AM/SSB-CW **Selectivity:** ±5.6 kHz
Circuit: Double Conversion **Physical:** 7.25x4.75x1.25" 1.5 Lbs.
Features: Mini Head. Jack, Tone Switch, 10 Memories, Sweep, Keypad, Fine Tuning, 12/24 Clock-Timer, Up-Down Tuning, Attenuator, LED Tune Indicator, 9/10 kHz MW Step, Record Output Jack, External Antenna Jack.
Accs.: AC-9W AC Adapter, DCC-127A Car Battery Cord.
New Price: $210-270 **Used Price:** $90-100 **Rating:** ★★★★★
Comments: Supplied with carry case, BP-25 external battery case (4xC cells) and earphone. Dark gray. German model **ICF-7600DS** covers 153-26100 kHz.

SONY®
ICF-2010

General Coverage Portable Communications Receiver
Made In: Japan 1985-1999 **Voltages:** 4.5 VDC
Coverage: 150-30000 kHz +FM +Air **Readout:** Digital LCD 0.1
Modes: AM/SSB-CW **Selectivity:** Two Position
Circuit: Double Conversion **Physical:** 11.4x6.25x2.2" 4 Lbs.
Features: Mini Head. Jack, Tone Switch, 32 Memories, Sweep, Keypad, 12/24 Clock-Timer, Up-Down Tuning, Dual Speed Manual Tuning, Attenuator, Synchronous Detection, 9/10 MW Step, Record Jack, LED Tune Indicator, Dial Lamp, Dial Lamp Switch, Ext. Ant. Jack.
Accs.: DCC-127A Car Battery Cord.
New Price: $350-450 **Used Price:** $220-230 **Rating:** ★★★★★
Comments: Supplied with an AC adapter and earphone. Operates from three D and two AA cells. This highly respected, long-lived model is a favorite among SW and MW DXer's. Also includes the VHF aeronautical band 116-136 MHz. Sold outside the U.S. as model **ICF-2001D**.

SONY®
ICF-6500W

General Coverage Portable Communications Receiver
Made In: Japan 1982-1985 **Voltages:** 120 VAC
Coverage: See Comments **Readout:** Digital LCD 5.
Modes: AM/SSB-CW **Selectivity:** ±9 kHz -60dB.
Circuit: Double Conversion **Physical:** 11.5x6.75x4.2" 4 Lbs.
Features: Mini Head. Jack, S-Meter, Tone Control, Sensitivity Switch, Carry Strap, FM AFC, BFO, External Antenna Jack, Dial Lock, Dial Lamp, Dial Lamp Switch, Two Manual Tuning Rates.
Accs.: DCC-120 Car Adapter
New Price: $150-200 **Used Price:** $70-80 **Rating:** ★★★★
Comments: Ranges: .53-1.605, 3.9-10, 11.7-20 and 20-28 MHz plus FM. Supplied with AC-160W AC adapter or operates from six C cells. Model **ICF-6500L** includes longwave.

SONY®
ICF-6700W

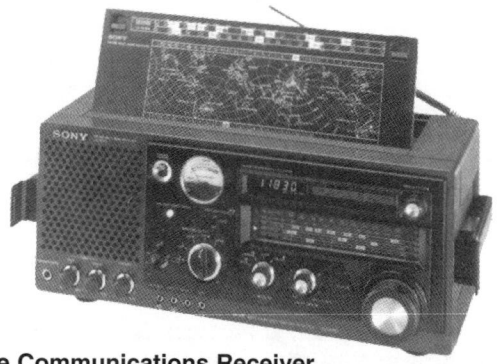

General Coverage Portable Communications Receiver
Made In: Japan 1978-1980 **Voltages:** 110/120/220/240 VAC
Coverage: 530-30000 kHz +FM **Readout:** Digital LED 1.
Modes: AM/USB/LSB-CW **Selectivity:** 10/4 kHz
Circuit: Double Conversion **Physical:** 18x7.5x9.25" 12 Lbs.
Features: ¼" Head. Jack, S/Battery Meter, Bass, Treble, Dial Lamp, Dial Lamp Switch, Preselector, FM AFC, Carry Strap, Timer Jack, Record Jack, Carry Handle, RF Gain, Time Chart, MPX Output.
Accs.: DCC-130 Car Battery Cord
New Price: $365-440 **Used Price:** $180-260 **Rating:** ★★★
Comments: Also operates from six D cells. Model **ICF-6700L** also covers longwave.

SONY.
ICF-6800W

General Coverage Portable Communications Receiver
Made In: Japan 1980-1983   **Voltages:**   110/120/220/240 VAC
Coverage: 530-30000 kHz +FM   **Readout:**   Digital LED 1.
Modes: AM/USB/LSB-CW   **Selectivity:** 9/4 kHz
Circuit: Double Conversion   **Physical:**   18x7.5x9.25" 12 Lbs.
Features: ¼" Head. Jack, S/Battery Meter, Bass, Treble, Dial Lamp,
               Dial Lamp Switch, Preselector, FM AFC, Carry Strap, Timer Jack,
               Record Jack, Carry Handle, Memo Lamp, Time Chart.
Accs.:   DCC-120 Car Battery Cord
New Price: $539-650   **Used Price:** $350-370   **Rating:** ★★★★
Comments: Also operates from six D cells. Model **ICF-6800WA Orange** is later production with improvements. Both models enjoy collector demand.

SONY.
ICF-SW1S

General Coverage Portable Broadcast Receiver
Made In: Japan 1988-1993   **Voltages:**   120 VAC or 2xAA
Coverage: 150-29995 kHz +FM^S **Readout:**   Digital LCD 5.
Modes: AM   **Selectivity:** One Position
Circuit: Double Conversion   **Physical:**   4.75x2.785x1" 0.5 Lbs.
Features: Mini Head. Jack, Tone, 10 Memories, Two Tuning Steps, Scan,
               Keypad, Clock-Timer, Sleep, Up-Down Tuning, Key Lock,
               9/10 kHz MW Step, Local-DX Switch, Dial Lamp, Tune LED.
Accs.:   DCC-E127A Car Battery Cord
New Price: $300-350   **Used Price:** $150-160   **Rating:** ★★★★
Comments: An amazing 4.375x1x2.9" at 8 oz. Supplied with AC adapter, stereo earphones, active antenna and carry case. Model **ICF-SW1E** is identical to the ICF-SW1S sans carrying case and most accessories. The ICF-SW1E was not sold in North America.

SONY®
ICF-SW55

General Coverage Portable Communications Receiver
Made In: Japan 1993-1998 **Voltages:** 120 VAC or 4xAA
Coverage: 150-30000 kHz +FM^S **Readout:** Digital LCD 1.
Modes: AM/USB/LSB **Selectivity:** Two Position
Circuit: Double Conversion **Physical:** 7.75x5x1.5" 1.8 Lbs.
Features: Mini Head. Jack, S-Indicator, Tone, 125 Alpha Memories, Sleep, RF Gain, Record Jack, Dial Lamp, Dial Lamp Switch, Tilt Bar, Two Tuning Rates, Keypad, Clock-Timer, Manual Tuning, Sweep, Up-Down Tune, Dial & Keypad Lock, Record Activation.
Accs.: DCC-E160L Car Battery Cord
New Price: $350-450 **Used Price:** $200-220 **Rating:** ★★★★
Comments: Surprisingly good audio for its compact size. Supplied with AC adapter stereo earphones, system carry case, windup antenna and external antenna connector.

SONY®
ICF-SW77

General Coverage Portable Communications Receiver
Made In: Japan 1992-1999 **Voltages:** 120 VAC
Coverage: 150-30000 kHz +FM^S **Readout:** Digital LCD 0.1
Modes: AM/USB/LSB/CW **Selectivity:** Two Position
Circuit: Double Conversion **Physical:** 11x7x2" 3.25 Lbs.
Features: Mini Head. Jack, S/Batt.-Indicator, Tone, 162 Alpha Memories, RF Gain, Record Jack, Dial Lamp, Dial Lamp Switch, Tilt Bar, Two Tuning Rates, Keypad, Synchronous Detection, Sleep, Clock-Timer, 9/10 kHz MW Step, Manual Tuning, Sweep, Up-Down Tuning, LCD Contrast Adjustment, Bass, Treble.
Accs.: DCC-E160L Car Battery Cord
New Price: $470-650 **Used Price:** $280-300 **Rating:** ★★★★★
Comments: Operates from four C cells. Supplied with AC adapter, carry belt, windup antenna and stereo earphones. Sony's flagship model.

SONY®
ICF-SW100S

General Coverage Portable Communications Receiver
Made In: Japan 1993-1999   **Voltages:** 120 VAC or 2xAA
Coverage: 150-30000 kHz +FM^S **Readout:** Digital LCD 1.
Modes: AM/USB/LSB-CW   **Selectivity:** One Position
Circuit: Double Conversion   **Physical:** 4.375x2.9x1" 0.5 Lbs.
Features: Mini Head. Jack, LED Tune Indicator, Tone, 50 Alpha Memories, Dial Lamp, Dial Lamp Switch, Scan, Sleep, Local-DX Switch, Two Tuning Steps, Keypad, Clock-Timer, Up-Down Tuning, Synchronous Det., Record Jack, 9/10 kHz MW Step, Key Lock.
Accs.: DCC-E130L Car Battery Cord
New Price: $359-450   **Used Price:** $220-230   **Rating:** ★★★★
Comments: An amazing 4.375x1x2.9" at 8 oz. Surprisingly good audio for its very compact size. Supplied with AC adapter stereo earphones, active antenna and carry case. The model **ICF-SW100E**, sold outside North America includes only the earphones, case and a nonactive active antenna. The predecessor model ICF-SW1S did not include Sync. Detection or feature SSB reception.

SONY®
ICF-SW1000T

General Coverage Portable Communications Receiver
Made In: Japan 1996-1999   **Voltages:** 3 VDC
Coverage: 150-30000 kHz +FM^S **Readout:** Digital LCD 1.
Modes: AM/USB/LSB-CW   **Selectivity:** One Position
Circuit: Double Conversion   **Physical:** 7x4.25x1.62" 1.25 Lbs.
Features: Mini Head. Jack, LED Tune Indicator, Tone Switch, 32 Memories, Dial Lamp, Dial Lamp Switch, Scan, Sleep, Local-DX Switch, Two Tuning Steps, Keypad, 24 Hour Clock-Timer, Mic Jack, Synchronous Detection, Record Jack, Key Lock, Up-Down Tuning, Auto-Reverse Stereo Cassette Deck.
Accs.: AC-E30HG AC Adapter
New Price: $449-599   **Used Price:** $180-190   **Rating:** ★★★★
Comments: Also operates from three AA cells. The integral stereo cassette is accessed from the back of the radio. Supplied with wind-up antenna, wrist strap, vinyl carrying case and stereo earphones. Model **ICF-SW1000TS** (new $500-650) includes the AC adapter and the Sony AN-LP1 active loop antenna.

SONY®
ICF-SW7600

General Coverage Portable Communications Receiver
Made In: Japan 1992-1993 **Voltages:** 6 VDC
Coverage: 150-30000 kHz +FM^S **Readout:** Digital LCD 5.
Modes: AM/LSB/USB **Selectivity:** One Position
Circuit: Double Conversion **Physical:** 7.5x4.75x1.3" 1.5 Lbs.
Features: Mini Head. Jack, Tone Switch, 10 Memories, Sleep, Sweep, Record Jack, Dial Lock, Keypad Lock, Local-DX Switch, LED Tune Indicator, Dual Clock Timers, 9/10 kHz MW Step, Dial Lamp, Dial Lamp Switch, External Antenna Jack.
Accs.: DCC-127A Car Battery Cord
New Price: $220-250   **Used Price:** $80-90   **Rating:** ★★★★★
Comments: Supplied with AC adapter, stereo earphones, wrist strap and AN-61 windup antenna. Operates from four AA cells. Gray.

SONY®
ICF-SW7600G

General Coverage Portable Communications Receiver
Made In: Japan 1994-1999 **Voltages:** 6 VDC
Coverage: 150-30000 kHz +FM^S **Readout:** Digital LCD 1.
Modes: AM/LSB/USB **Selectivity:** One Position
Circuit: Double Conversion **Physical:** 7.5x4.75x1.3" 1.5 Lbs.
Features: Mini Head. Jack, Tone Switch, 22 Memories, Sweep, Rec. Jack, Fine Tuning, Clock-Timer, Up-Down Tune, DX-Local, Dial Lock, LED Tune Indicator, 9/10 kHz MW Step, Ext. Ant. Jack, Keypad, Dial Lamp, Dial Lamp Switch, Sleep, Synchronous Detection.
Accs.: E60HG/AC60M AC Adapter, DCC-E260HG Car Battery Cord
New Price: $170-230   **Used Price:** $100-105   **Rating:** ★★★★★
Comments: Supplied with vinyl case and windup antenna. Operates from four AA cells. The least expensive receiver to feature synchronous detection. The model **ICF-SW7600GS** "system" includes the Sony AN-LP1 active loop antenna and Sony AC adapter ($250-290 new).

19 Yaesu

Yaesu U.S.A.
17210 Edwards Rd.
Cerritos, CA 90701

YAESU FRG-7

General Coverage Communications Receiver
Made In: Japan 1976-1980 **Voltages:** 100/110/117/200/220
Coverage: 500-29900 kHz **Readout:** Analog Linear
Modes: AM/USB/LSB-CW **Selectivity:** 6 kHz
Circuit: Triple Conversion **Physical:** 13.5x6.5x11.5" 16 Lbs.
Features: ¼" Head. Jack, S-Meter, Attenuator, Tone Switch, Mute Line, Fine Tuning, Record Jack, ANL, Dial Adj., Dial Lamp Switch.
Accs.: DC Kit, Internal Battery Holder (8xD).
New Price: $290-370   **Used Price:** $160-250   **Rating:** ★★★★★
Comments: The Fine Tuning knob is not featured in very early production. A great value on the used market.

YAESU FRG-100

General Coverage Communications Receiver
Made In: Japan 1994-1999 **Voltages:** 120 VAC 12 VDC
Coverage: 50-30000 kHz **Readout:** Digital LCD .01
Modes: AM/USB/LSB-CW **Selectivity:** 6/4/2.4/_ kHz
Circuit: Double Conversion **Physical:** 9.37x3.6x9.5" 6.6 Lbs.
Features: ¼" Head. Jack, S-Meter, Atten., Mute Line, ANL, AGC, Dimmer, Clock Timer, Dial Lock, Scan, Sweep, 50 Mems., Squelch, AGC, Carry Handle, 3 Tune Rates, RF Gain, Backlit LCD, Remote Jack.
Accs.: YF-100C 500 Hz Filter, YF-100CN 300 Hz Filter, FIF CAT Interface, FM-100 FM Unit, TCXO-4 High Stability Option.
New Price: $590-650   **Used Price:** $320-400   **Rating:** ★★★★★
Comments: This receiver has many user programmable features. A third-party wired keypad is available manufactured by B.E.E.I. of France ($60).

YAESU
FRG-7000

General Coverage Communications Receiver
Made In: Japan 1977-1980 **Voltages:** 100/110/117/200/220/234
Coverage: 250-29900 kHz **Readout:** Digital LED 1.
Modes: AM/USB/LSB-CW **Selectivity:** 6/3 kHz
Circuit: Triple Conversion **Physical:** 14.3x5x11.5" 16 Lbs.
Features: ¼" Head. Jack, S-Meter, Attenuator, Tone Control, Mute Line, Digital Clock Timer, Carry Handle, ANL, Record Jack, RF Gain, Recorder Activation Jacks, Fine Tuning, Preselector, Digital Display On-Off Switch.
New Price: $599-655 **Used Price:** $180-280 **Rating:** ★★★
Comments: This successor to the famous FRG-7 afforded digital readout, but less performance.

YAESU
FRG-7700

General Coverage Communications Receiver
Made In: Japan 1981-1984 **Voltages:** 100/120/200/240 VAC
Coverage: 150-29999 kHz **Readout:** Digital LED 1.
Modes: AM/USB/LSB-CW/FM **Selectivity:** 12/6/2.7 kHz
Circuit: Triple Conversion **Physical:** 13x5x10" 15 Lbs.
Features: ¼" Head. Jack, S-Meter, Attenuator, Tone, Mute Line, ANL, Clock Timer, Record Jack, Carry Handle, AGC, FM Squelch, Fine Tuning (for optional memory).
Accs.: MU-7700 12 Channel Memory, FRA-7700 Active Antenna, FRT-7700 Tuner, FF-5 Low Pass Filter, DC-7700 DC Kit, FRV-7700 External VHF Converter (six versions).
New Price: $400-550 **Used Price:** $260-340 **Rating:** ★★★
Comments: The optional MU-7700 memory unit mounts internally without special tools.

YAESU
FRG-8800

General Coverage Communications Receiver
Made In: Japan 1985-1993 **Voltages:** 100/120/200/240 VAC
Coverage: 150-29999 kHz **Readout:** Digital LCD 0.1
Modes: AM/USB/LSB-CW/FM **Selectivity:** 6/2.7 kHz [12.5 FM]
Circuit: Double Conversion **Physical:** 13x5x8.75" 13.2 Lbs.
Features: ¼" Head. Jack, S-Indicator, Atten., Tone, Mute Line, CAT Jack, Dual Clock Timer, Record Jack, Handle, AGC, Squelch, NB, Fine Tuning, Keypad, 12 Memories, Line Out, Two Tune Rates.
Accs.: FIF Cat Interface, FRV-8800 VHF Converter, FRT-7700 Tuner, DC-8800 DC Kit, FRA-7700 Active Antenna.
New Price: $429-699 **Used Price:** $400-430 **Rating:** ★★★
Comments: Requires three AA cells for the memory.

20 Other Manufacturers

Bearcat
DX-1000

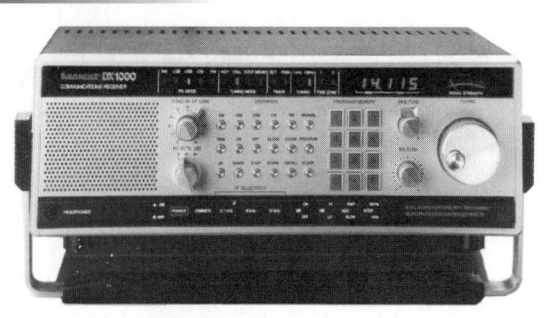

General Coverage Communications Receiver
Made In:	Japan 1983-1984	**Voltages:**	120/240 VAC 12 VDC
Coverage:	10-30000 kHz	**Readout:**	Digital LED 1.
Modes:	AM/USB/LSB/CW/FM	**Selectivity:**	12/6/2.7 kHz
Circuit:	Double Conversion	**Physical:**	14.5x5x9" 17.6 Lbs.

Features: ¼" Head. Jack, S-Meter, Attenuator, Tone, Mute Line, NB, AGC, Fine Tuning, Record Jack, Dual 24 Hr. Clock, Squelch, Dimmer, Two Tuning Rates, 10 Memories, Keypad, Tilt Carry Handle.
New Price: $500-550   **Used Price:** $150-250   **Rating:** ★
Comments: This receiver did not live up to the manufacturer's specifications. It was rich in features, but poor in performance. Also operates from 8 D cells.

GALAXY
R-530

General Coverage Communications Receiver
Made In:	United States 1967-1973	**Voltages:**	115/230 VAC
Coverage:	500-30000 kHz	**Readout:**	Analog Linear
Modes:	AM/USB/LSB	**Selectivity:**	2.1/_/_ kHz
Circuit:	Double Conversion	**Physical:**	17x6x14" 25 Lbs.

Features: ¼" Head. Jack, S/AF Meter, Attenuator, Mute Line, AVC, NB, Line Output Terminals, VFO Input/Output Jack, Preselector, BFO, RF Gain, Standby, Calibrator, AGC Output Jack.
Accs: SPK530 Speaker, CL530 Clock (for speaker), FL5305 500 Hz CW Filter, FL5306 6 kHz AM Filter, FL53015 1.5 kHz Filter.
New Price: $695-895   **Used Price:** $490-650   **Rating:** ★★★★
Comments: A very advanced and accurate receiver for its time. Dial accuracy was ±1 kHz. Very scarce. Caution: This company is no longer in business. The model **R-1530** is a military version.

RACAL
COMMUNICATIONS, INC.
RA6790/GM

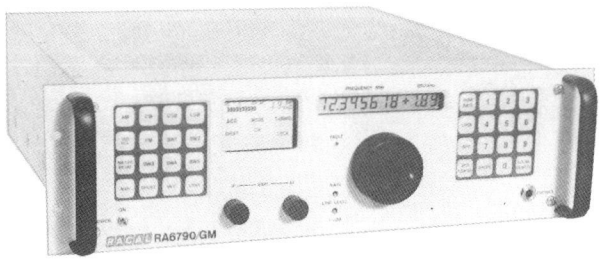

General Coverage Communications Receiver
Made In: United States 1979-1988   **Voltages:**   110/120/200/240 VAC
Coverage: 500-30000 kHz   **Readout:**   Digital LCD 0.001
Modes: AM/LSB/USB/CW/FM   **Selectivity:** 20/6/3.2/1/.3 kHz
Circuit: Double Conversion   **Physical:**   19x5.25x18.5" 32 Lbs.
Features: ¼" Head. Jack, S/AF Indicator, Keypad, 3 Tuning Rates, AGC, Mute, Rack Handles, IF Out Jack, Lock, BFO, BITE, RF Gain, Line Out Jacks (2), External Speaker Jack.
Accs.:   LF Extension, VLF Extension, I.F. Converter, Very NBFM, MA6004 Control Unit, ISB 2 Channel, ISB 4 Channel, Ext. Ref. Jack, High Stability Option, Serial Port.
New Price: $5000-6000   **Used Price:** $780-1600   **Rating:** ★★★★★
Comments: The earlier model **RA6790** is similar, but without LED on the front panel for Fault. Often sold on the surplus market without *any* I.F. filters.

TEN-TEC
RX-325

General Coverage Communications Receiver
Made In: United States 1987-1988   **Voltages:**   115 VAC 13.8 VDC
Coverage: 300-30000 kHz   **Readout:**   Digital LED 0.1
Modes: AM/USB/LSB   **Selectivity:** 4/2.7/_ kHz
Circuit: Double Conversion   **Physical:**   9.5x3.25x7" 6 Lbs.
Features: Mini Head. Jack, S-Meter, Attenuator, Dial Lock, NB, Sweep, 25 Memories, Memory, Keypad, AGC, 12/24 Hour Clock, Scan, Sweep, Dimmer, Two Tuning Speeds, Two Tuning Rates, Record Activation.
Accs: 265 Speaker, 266 SSB Filter, 267 Timer Relay, 925 Battery Pack.
New Price: $549-629   **Used Price:** $275-350   **Rating:** ★
Comments: Supplied with external AC adapter. Scarce.

uniden
CR-2021

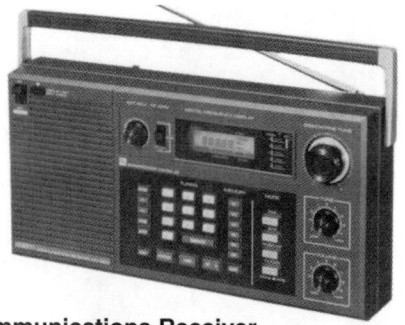

General Coverage Portable Communications Receiver
Made In:	Taiwan 1983-1984	**Voltages:**	120 VAC 11-16 VDC
Coverage:	150-30000 kHz +FM	**Readout:**	Digital LCD 1.
Modes:	AM/SSB-CW	**Selectivity:**	3/2.5 kHz
Circuit:	Triple Conversion	**Physical:**	13x6.5x3.5" 5 Lbs.

Features: ¼" Head. Jack, S/Battery Indicator, Attenuator, BFO, Tone, 3 Position RF Gain, Sleep, Two Tuning Steps, Carry Handle, Record Jack, 24 Hour Clock, Sweep, Antenna Trimmer, Keypad, 12 Memories, Dial Lamp, Dial Lamp Switch, Antenna Terminals.

New Price: $150-270   **Used Price:** $85-90   **Rating:** ★★★

Comments: The CR-2021 can store six FM stations and six non-FM stations for a total of twelve. Operates from six C cells and two AA cells. The **Realistic DX-400** is similar, but features an analog S-Meter rather than a LED S-Indicator.

WATKINS-JOHNSON
HF1000

General Coverage Communications Receiver
Made In:	United States 1993-1999	**Voltages:**	97-253 VAC
Coverage:	5-30000 kHz	**Readout:**	Digital LED 0.001
Modes:	AM/USB/LSB/ISB/FM/CW	**Selectivity:**	See Comments.
Circuit:	Double Conversion	**Physical:**	19x5.25x20" 15 Lbs.

Features: ¼" Head. Jack, S-Meter, Attenuator, DSP IF, NB, PBT, Preamp, 100 Memories, Memory Scan, Sweep, Keypad, AGC, Squelch, Channel Lockout, RS-232, BITE, IF Output, BFO, Rack Handles, Notch, Dial Lock, Sync. Detection, Mute, Speaker Switch.

Accs.: Internal Suboctave Preselector

New Price: $3800-4000   **Used Price:** $2700-2800   **Rating:** ★★★★★

Comments: Supplied bandwidths: 8000, 7200, 6400, 6000, 5600, 5200, 4800, 4400, 4000, 3600, 3200, 3000, 2800, 2600, 2400, 2200, 2000, 1800, 1600, 1500, 1400, 1300, 1200, 1100, 1000, 900, 800, 750, 700, 650, 600, 550, 450, 400, 375, 350, 325, 300, 275, 250, 225, 200, 188, 175, 163, 150, 138, 125, 113, 100, 94, 88, 81, 75, 69, 63 and 56 Hz -3dB. Scarce.

21 Model Index

A

AR3030	25
AR7030	25
AR7030+	25
ATS-803	60
ATS-803A	60
ATS-808	60
ATS-808A	60
ATS-818	61
ATS-818CS	61
ATS-909	61
AX-190	24

C

CR-2021	76
CRF-1	62
CRF-320A	62
CRF-330K	63
CRF-V21	63

D

D2935	45
D2999	45
DR22	46
DR22C-6	46
DR26	48
DR28	49
DR29	49
DR31	49
DR33C-6	46
DR44	47
DR44-6	47
DR101-6	47
DR-B300	54
DR-B600	54
DR Q 63	51
DX-100	55
DX-120	55
DX-150	56
DX-150A	56
DX-150B	57
DX-160	57
DX-200	58
DX-300	58
DX-302	59
DX-380	60
DX-390	61
DX-392	61
DX-394	59
DX-398	61
DX-400	75
DX-440	60
DX-1000	74

F

FRG-7	71
FRG-100	71
FRG-7000	72
FRG-7700	72
FRG-8800	73

H

HF-125	43
HF-150	43
HF-150E	43
HF-150M	43
HF-225	44
HF-225E	44
HF-250	44
HF-250E	44
HF1000	76

I

ICF-2001	64
ICF-2001D	65
ICF-2002	64
ICF-2003	65
ICF-2010	65
ICF-6500L	66
ICF-6500W	66
ICF-6700L	66
ICF-6700W	66
ICF-6800W	67
ICF-6800WA	67
ICF-7600D	65

ICF-7600DS	65
ICF-SW1E	67
ICF-SW1S	67
ICF-SW55	68
ICF-SW77	68
ICF-SW100E	69
ICF-SW100S	69
ICF-SW1000T	69
ICF-SW1000TS	69
ICF-SW7600	70
ICF-SW7600D	64
ICF-SW7600DS	65
ICF-SW7600G	70
ICF-SW7600GS	70

N

NRD-345	37
NRD-505	37
NRD-515	38
NRD-525	38
NRD-535	39
NRD-535D	39
NRD-535V	39
NRD-545	39

P

PCR100	35
PCR1000	35
PCR1000-02	35
PRN1000	29

R

R7	26
R7A	26
R8	27
R8A	27
R8B	28
R-70	35
R-71A	36
R-72	36
R-300	40
R-530	74
R-600	40
R-1000	41
R-1530	74
R-2000	41
R4245	26
R-5000	42
RA6790	75
RA6790/GM	75
RF-799	48
RF-799LBE	48
RF-799LBS	48
RF-2600	48
RF-2800	49
RF-2900	49
RF-3100	49
RF-4800	50
RF-4900	50
RF-6300	51
RF-6300L	51
RF-9000	51
RF-B30	49
RF-B40	52
RF-B45	52
RF-B60	53
RF-B65	53
RF-B300	54
RF-B600	54
RX-325	75

S

Satellit 400	32
Satellit 500	32
Satellit 650	33
Satellit 700	33
SPR-4	28
SSR-1	29
SW1	29
SW2	30
SW-4	30
SW-4A	30
SW8	31
SX-190	24

Y

YB-400	34
YB-400PE	34
YB-500	34